16
Topics in Medicinal Chemistry

Aims and Scope

Drug research requires interdisciplinary team-work at the interface between chemistry, biology and medicine. Therefore, the new topic-related series Topics in Medicinal Chemistry will cover all relevant aspects of drug research, e.g. pathobiochemistry of diseases, identification and validation of (emerging) drug targets, structural biology, drugability of targets, drug design approaches, chemogenomics, synthetic chemistry including combinatorial methods, bioorganic chemistry, natural compounds, high-throughput screening, pharmacological in vitro and in vivo investigations, drug-receptor interactions on the molecular level, structure-activity relationships, drug absorption, distribution, metabolism, elimination, toxicology and pharmacogenomics.

In general, special volumes are edited by well known guest editors.

In references Topics in Medicinal Chemistry is abbreviated Top Med Chem and is cited as a journal.

More information about this series at
http://www.springer.com/series/7355

Martial Ruat
Editor

The Smoothened Receptor in Cancer and Regenerative Medicine

With contributions by

R. Andriantsitohaina · M. Daynac · H. Faure · L. Hoch · J. Jiang · F. Manetti · M.C. Martínez · I. Mus-Veteau · E. Petricci · N.A. Riobo · D. Rognan · M. Ruat · Q. Shi · R. Soleti · M. Taddei

Springer

Editor
Martial Ruat
Institute of Neurosciences Paris-Saclay (Neuro-PSI)
National Centre for Scientific Research
Gif-Yvette, France

ISSN 1862-2461 ISSN 1862-247X (electronic)
Topics in Medicinal Chemistry
ISBN 978-3-319-19754-8 ISBN 978-3-319-19755-5 (eBook)
DOI 10.1007/978-3-319-19755-5

Library of Congress Control Number: 2015943691

Springer Cham Heidelberg New York Dordrecht London

Printed on acid-free paper

Springer International Publishing AG Switzerland is part of Springer Science+Business Media (www.springer.com)

Preface

Smoothened (Smo) belongs to class F (Frizzled family) of the G-protein-coupled receptor (GPCR) superfamily. Smo is the main transducer of the Hedgehog (Hh) signaling pathway, which is responsible for a large variety of developmental processes. This receptor regulates key physiological responses in adult tissues, including the maintenance of stem and precursor cells. Recent clinical investigations implicate Smo as a novel therapeutic target in multiple human cancers.

There is evidence that Hh signaling is upregulated in a number of cancers. Thus, the discovery of the natural and teratogenic alkaloid molecule cyclopamine, which binds to the 7-transmembrane domain of Smo blocking canonical Hh signaling, heralded extensive research for Smo antagonist development. This led to the recent approval of GDC-0449 (vismodegib) for treating metastatic basal cell carcinoma (BCC) and locally advanced BCC, untreatable by surgery and radiation. The importance of Smo inhibitors, of different chemical classes, is also highlighted by ongoing clinical trials on a large range of metastatic and advanced cancers.

In this volume of *Topics in Medicinal Chemistry* focused on Smo, the authors cover the most recent data describing the key role of this receptor as a major component of the Hh signaling pathway in vertebrates. They summarize major advances recently achieved on Hh pathway activation with a special focus on Smo signaling mechanisms and regulation. The authors emphasize the intense academic and clinical efforts to translate basic developmental biology research into the clinic.

The first chapter by Lucile Hoch and myself (CNRS, Neuroscience Paris-Saclay Institute, University Paris-Sud, France) reviews recent preclinical and clinical data on the therapeutic importance of Smo and highlights the complexity of Smo pharmacology and its clinical implications. In the second chapter, Natalia A. Riobo from the Kimmel Cancer Center at Thomas Jefferson University (Philadelphia, USA) describes in detail the roles and functions of Hh pathway components in canonical and noncanonical Hh signaling. The role of Smo as a GPCR that selectively activates heterotrimeric G proteins is presented and discussed in the context of cancer, angiogenesis, and fibrosis.

Fabrizio Manetti and his colleagues, Elena Petricci and Maurizio Taddei at the University of Siena (Italy), provide an accurate update of synthetic and natural molecules modulating the Hh signaling pathway. This extensive review will be of invaluable help to students, medicinal chemists, and biologists interested in structure–activity relationships and mechanisms of action of both Smo agonists and antagonists, including positive and negative allosteric modulators. In their review of the most recent data on how Hh proteins dynamically induce Smo phosphorylation and ubiquitination, Qing Shi and Jin Jiang from the Texas Southwestern Medical Center, University of Texas (Dallas, USA) discuss how these modifications modulate Smo function and downstream signaling events. This review should impact future exploration of the mechanisms of action of drugs interacting with Smo.

Understanding the binding mode of Smo antagonists is of prime importance to delineate the complex processes that induce Smo resistance to drug treatment. Didier Rognan (CNRS, University of Strasbourg, France) and Isabelle Mus-Veteau (CNRS, University of Nice-Sophia Antipolis, France) discuss the very plastic nature of Smo, which is able to accommodate a diverse array of ligands through several binding sites, highlighted by the recent X-ray structures of the transmembrane and the extracellular domains of the human Smo receptor. These structural data may help with designing novel molecules that can act at different Smo binding sites and are clinically insensitive to Smo mutations.

The chapter by Mathieu Daynac, Hélène Faure, and myself discusses the emerging role of Smo in the adult brain, particularly in the maintenance of neural stem cells. We unravel research on the clinical implications for the treatment of brain diseases, including Hh-linked medulloblastoma. In the last chapter, Raffaella Soleti, Ramaroson Andriantsitohaina, and Maria Carmen Martínez (INSERM, Hospital and University Center of Angers, France) describe the important role Hh signaling plays in neovascularization, which represents a key step in tissue repair. They discuss both in vitro and in vivo data that support the therapeutic potential of the Hh pathway and Smo for myocardial infarction and diabetes.

Despite strong evidence for the therapeutic potential of targeting the Hh pathway in cancer therapy, we need to understand more about Smo in tumorigenesis. Smo agonists may have therapeutic value in central nervous diseases, ischemia or diabetes, but are yet to be developed for clinical use. Thus, this volume would be suitable not only for students but also for chemists, biologists, and clinical scientists in understanding how Smo activity can be regulated in tissues and identifying the potential therapeutic effects of Smo modulators.

I would like to express my deep appreciation to the contributing authors of this special volume for their excellent and deep analyses of the current status of the roles of Smo in tissues and its regulation by small molecules for therapeutic applications. I also deeply thank the series editors for their enthusiasm in publishing this volume.

Gif-sur-Yvette, France Martial Ruat

Contents

Contents

Top Med Chem (2015) 16: 1–12
DOI: 10.1007/7355_2015_85

Published online: 17 April 2015

Smoothened Inhibitors in Cancer

Martial Ruat and Lucile Hoch

Abstract Smoothened (Smo) inhibitors are under intense development for the treatment of cancers linked to abnormal Hedgehog (Hh) signaling. The first inhibitor (vismodegib) was introduced in clinics for basal cell carcinoma and medulloblastomas associated with activating mutations of Hh signaling. In contrast, disappointing data are reported for cancers related to ligand overexpression. Here, we review recent preclinical and clinical data on the potential therapeutic importance of Smo and highlight the complexity of Smo pharmacology and its clinical implications.

Keywords Cancer stem cell, Hedgehog, Medulloblastoma, Resistance

Contents

M. Ruat (✉) and L. Hoch
Signal Transduction and Developmental Neuropharmacology Team, Department of Molecules and Circuits, CNRS, UMR 9197, Neurosciences Paris-Saclay Institute, Paris Sud University, 1 Avenue de la Terrasse, F-91198 Gif-sur-Yvette, France
e-mail: ruat@inaf.cnrs-gif.fr

Abbreviations

7TM	7-Transmembrane domain
BCC	Basal cell carcinoma
Gli1–3	Glioma-associated oncogenes 1–3
GPCR	G-protein-coupled receptor
Hh	Hedgehog
PKA	Protein kinase A
Ptc	Patched
Smo	Smoothened
Sufu	Suppressor of fused

1 Introduction

Smoothened (Smo), a member of the G-protein-coupled receptor (GPCR) superfamily, is the main transducer of the Hedgehog (Hh) signaling pathway. This pathway is implicated in the maintenance of stem cells and tissue repair in the adult. However, aberrant control of this pathway is associated with tumorigenesis. Thus, intense academic and clinical research has focused on designing potent Smo inhibitors and determining their functionality for manipulating Smo activity in various cancers. A major breakthrough in the Hh field is the recent approval of Erivedge/vismodegib (GDC-0449, Genentech, Figure 1) by the FDA for treating metastatic basal cell carcinoma (BCC) and locally advanced BCC untreatable by surgery or radiation. Several clinical trials are underway in which Smo inhibitors are combined with other therapeutics for the treatment of a wide variety of solid tumors and blood malignancies [1–6]. Here, we discuss recent findings on Hh pathway activation and data from various clinical trials with Smo inhibitors targeted to blocking Hh signaling in cancer.

2 Smoothened, a Therapeutic Target for Cancer Therapy

2.1 Transduction of the Hedgehog Signal

In the absence of Hh ligands, the 12-pass transmembrane protein Patched (Ptc) negatively regulates Smo presumably via transporter-like activity. The binding of Hh to Ptc activates the canonical Hh signaling pathway by translocating Smo to the primary cilium. This initiates a complex signaling cascade mediated by the activation of the zinc finger transcription factors, glioma-associated oncogenes 1–3 (Gli1, Gli2, and Gli3), and translocation of their active forms to the nucleus leading to gene transcription [2, 7]. Interestingly, the primary cilium recently emerged as an important center for Hh pathway transduction in vertebrates (Figure 2). Trafficking

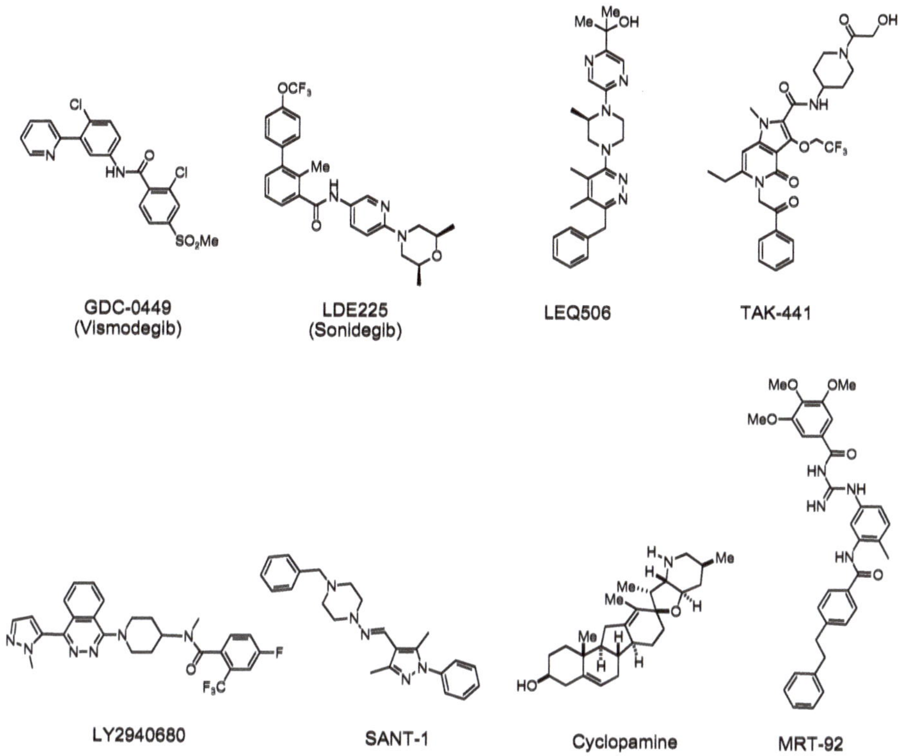

Fig. 1 Chemical structures of Smo antagonists

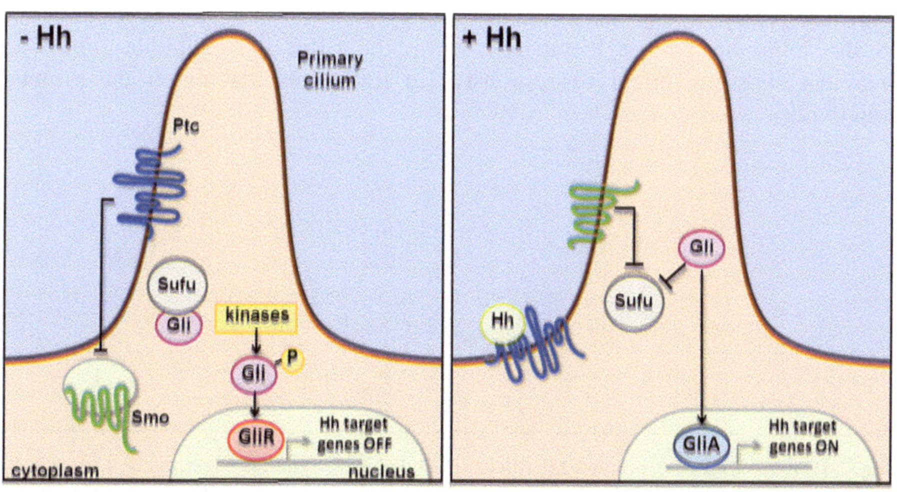

Fig. 2 Hedgehog signaling pathway at the primary cilium. In the absence of Hedgehog ligand (-Hh), the receptor Patched (Ptc), located in the cilium, inhibits Smoothened (Smo), a 7-transmembrane receptor mostly found outside the cilium, by a yet unknown mechanism. Repressor factors such as suppressor of fused (Sufu) and kinases, including protein kinase A

of Hh signaling proteins, along the cilia of stem and precursor cells, is the key step in the neural development of several genetic diseases and cancer (reviewed in [4, 8–10]).

2.2 Ligand-Independent Hh-Associated Cancers

The Hh signaling pathway is associated to cancer development due to the identification of germline loss-of-function *Ptc* mutations in patients with Gorlin syndrome (or nevoid basal cell carcinoma syndrome), an autosomal dominant disease [11]. These patients are predisposed to developing medulloblastoma, BCC, rhabdomyosarcoma, meningioma, as well as tumors localized to the jaw [1, 12, 13]. Somatic mutations of *Ptc* and *Smo* were identified in sporadic BCC and medulloblastomas [14–19]. Somatic gain-of-function mutations of *Smo* are also reported in meningiomas [20, 21] and are believed to increase tumorigenesis through the aberrant activation of Hh signaling [18, 22]. Similarly, somatic and germline mutations in *Suppressor of fused* (*Sufu*) are associated with medulloblastoma [23, 24]. The alteration of Hh signaling due to *Ptc* mutations was recently identified using an integrative deep-sequencing analysis of children with medulloblastoma [25–27]. Somatic *Ptc* mutations were also identified in ovarian and endometrial cancers, but their association with neoplasia requires further investigation [1].

Interestingly, *Ptc* heterozygous mice develop cerebellar tumors resembling human medulloblastoma. These tumor cells were used to develop a mouse model for investigating the potency of Smo inhibitors in blocking tumor progression [28]. The presence of primary cilia in specific variants of human medulloblastoma is also important from a therapeutic viewpoint. Ciliated medulloblastoma with high Hh signaling might be responsive to treatments that target the primary cilium [29].

Fig. 2 (continued) (PKA), promote Gli truncation and phosphorylation, respectively. These events lead to the generation of Gli repressor forms (GliR) and inhibition of Hh target genes. In the presence of Hh ligand (+Hh), Smo inhibition is relieved allowing its translocation and accumulation in the cilium. This leads to Sufu inhibition and Gli conversion into their activated forms (GliA). GliA enters the nucleus and activates transcription of Hh target genes including Ptc and Gli1

2.3 Ligand-Dependent Hh-Associated Cancers

Hh signaling is proposed to be responsible for the development of a variety of tumors through autocrine or paracrine ligand-dependent mechanisms. Secretion of one of the Hh peptides (Sonic, Indian, or Desert Hedgehog) from either the tumor or the stromal environment is implicated in the transformed phenotype. These tumors are called ligand dependent, and several Hh pathway activity models were discovered with therapeutic implications [3, 4, 30]. The pathway was also associated to blood malignancies. Several recent reviews describe the process of autocrine or paracrine Hh pathway activation in different cancers, in detail [1, 4, 28, 31]. Pharmacological treatment using Smo inhibitors (Figure 1) in mouse models of these cancers is also reported. However, conflicting views on the molecular mechanisms of action in tumor regression exist, including the potential off-target effects of some of these drugs [30].

2.4 Development of Smoothened Antagonists for Cancer Therapy

Cyclopamine, a natural and teratogenic alkaloid molecule, which can be purified from corn lilies, slows down tumor growth in animal models [32–35]. This molecule, which is well known for inducing cyclopia in newborn sheep, blocks canonical Hh signaling presumably by binding to the 7-transmembrane (7TM) domain of Smo [36, 37]. Cyclopamine was not developed for therapeutic use, but a more soluble and potent derivative (IPI-926, saridegib) has entered clinical trials for treating BCC and metastatic pancreatic cancer [38]. Several potent Smo inhibitors of different chemical classes were developed in recent years, both by academia and the pharmaceutical industry. Several of these molecules demonstrated efficacy in mouse xenografts, leading to clinical trials on a large range of metastatic and advanced cancers [1, 30, 32–35, 39, 40]. Data from five clinical trials suggest that Smo inhibitor side effects include hair loss, muscle spasms, taste disturbance, fatigue, nausea, and decrease in weight and appetite [1]. Although these side effects are often moderate, they may result in treatment interruption in patients.

This extensive research led to the recent approval of GDC-0449 for treating BCC and locally advanced BCC, untreatable by surgery and radiation [41, 42]. Vismodegib is now authorized in several countries including the Europe Union, Australia, and South Korea for treating metastatic and locally advanced BCC [5].

Smo inhibition by vismodegib blocks the transcription of tumor mediating genes associated to the Hh pathway [43, 44]. However, a patient with a metastatic form of medulloblastoma had a relapse after initially responding to the drug. This was due to an Smo mutation (D473H$^{6.55}$) in the sixth transmembrane domain that disrupted vismodegib binding [45]. Likewise, a mutation occurring at a homologous position in mouse Smo was also observed in a vismodegib-resistant mouse model for medulloblastoma [46]. Furthermore, acquired resistance was also reported in BCC patients under vismodegib treatment [47–49]. Several Smo inhibitors such as sonidegib (LDE225, Novartis, Figure 1), BMS-833923, and saridegib are effective in BCC treatment and might be useful for treating Smo resistance. BCC patients, who had improved following vismodegib treatment, demonstrated limited benefit from saridegib. These results suggest an overlapping resistance mechanism, which is not yet investigated [38]. Analysis of resistance mechanisms in a medulloblastoma mouse model treated with LDE225 demonstrated activating Smo mutations, phosphatidylinositol 3-kinase upregulation, and *Gli2* amplification [50]. This suggests that besides resistance at the level of Smo itself, downstream Hh target gene amplification might also have clinical relevance. Thus, monitoring and establishing the resistance mechanisms associated with Smo inhibitor treatments in the ongoing trials will be important.

Sonidegib is currently in Phase III trials on medulloblastoma patients, selected for Smo inhibitor therapy based on a five-gene Hh signature [1]. Such an approach is expected to increase the number of patients positively responding to the treatment.

Smo inhibitors are being investigated in clinical trials on a variety of ligand-dependent tumors, but results are less successful (see also trial numbers NCT00822458, NCT01601184, NCT01239316, NCT01125800, NCT01208831, and NCT00880308, at *ClinicalTrials.gov*). Most of these trials are evaluating the effects of an Smo inhibitor with other therapeutic modalities. For example, treatment of metastatic pancreatic cancer patients with saridegib and gemcitabine revealed shorter median survival compared to those treated with gemcitabine alone [51]. Saridegib and TAK-441 have been discontinued, and negative results were obtained with vismodegib in patients with metastatic colorectal carcinoma and ovarian cancers [51–53]. It is important to understand the reasons for these failures, which might be linked to the mode of action of these compounds on Smo, Hh activation mechanisms in the tumor and trial design.

The antifungal compound itraconazole, an FDA-approved drug, inhibits Hh signaling and delays tumor growth, presumably by binding to hSmo at a site different from that of cyclopamine [54, 55]. This molecule is under investigation for BCC treatment [56].

3 Future Directions

X-ray structures of hSmo bound to several ligands have revealed two types of 7TM-directed antagonists: those binding mostly to extracellular loops (site 1, e.g., LY2940680) and those deeply penetrating the 7TM cavity (site 2, e.g., SANT-1). However, the existence of a third type of Smo antagonist was recently demonstrated. This class entirely fills the Smo binding cavity from the upper extracellular part to the lower cytoplasmic-proximal subpocket. One of these Smo inhibitors is the acylguanidine, MRT-92, which was shown to inhibit the Hh canonical signaling pathway and rodent cerebellar granule cell proliferation induced by Hh pathway activation [55]. MRT-92 is one of the most potent Smo inhibitors known to date and displays low sensitivity to block the effects of Smo conformational states associated to noncanonical pathways [55, 57]. To better understand the failures of Smo inhibitors observed in the clinic, it is of utmost importance to relate the clinical efficacy of Smo antagonists to their binding mode and to check whether a highly potent type 3 antagonist like MRT-92 may confer some advantages over the existing type 1 or type 2 Smo antagonists.

The identification of canonical and noncanonical pathways mediated through Smo culminates with the hypothesis that Smo antagonists of one pathway can act as agonists in another pathway [2, 58, 59]. This is reminiscent of the signaling bias reported for an increasing number of molecules acting on GPCRs [60]. The recognition that several agonists do not stabilize the same active site, but rather unique active states of a given receptor, fits with most Smo modulator pharmacological data. Smo interacts with Gi family members, presumably to decrease cAMP levels [61, 62], and might be associated with multiple cellular signaling proteins [2]. Thus, it will be important to better understand Smo regulation by small molecules, biased signaling, and associated pathways, which should help identify potential therapeutic effects of Smo modulators.

References

1. Amakye D, Jagani Z, Dorsch M (2013) Unraveling the therapeutic potential of the hedgehog pathway in cancer. Nat Med 19:1410–1422
2. Ruat M, Hoch L, Faure H, Rognan D (2014) Targeting of smoothened for therapeutic gain. Trends Pharmacol Sci 35:237–246
3. Heretsch P, Tzagkaroulaki L, Giannis A (2010) Modulators of the hedgehog signaling pathway. Bioorg Med Chem 18:6613–6624
4. Scales SJ, de Sauvage FJ (2009) Mechanisms of hedgehog pathway activation in cancer and implications for therapy. Trends Pharmacol Sci 30:303–312
5. Basset-Seguin N, Sharpe HJ, de Sauvage FJ (2015) Efficacy of hedgehog pathway inhibitors in basal cell carcinoma. Mol Cancer Ther
6. Justilien V, Fields AP (2015) Molecular pathways: novel approaches for improved therapeutic targeting of hedgehog signaling in cancer stem cells. Clin Cancer Res 21:505–513
7. Briscoe J, Therond PP (2013) The mechanisms of hedgehog signalling and its roles in development and disease. Nat Rev 14:416–429
8. Ruat M, Roudaut H, Ferent J, Traiffort E (2012) Hedgehog trafficking, cilia and brain functions. Differentiation 83:S97–S104
9. Goetz SC, Anderson KV (2010) The primary cilium: a signalling centre during vertebrate development. Nat Rev Genet 11:331–344
10. Louvi A, Grove EA (2011) Cilia in the CNS: the quiet organelle claims center stage. Neuron 69:1046–1060
11. Hahn H, Wicking C, Zaphiropoulos PG, Gailani MR, Shanley S, Chidambaram A, Vorechovsky I, Holmberg E, Unden AB, Gillies S, Negus K, Smyth I, Pressman C, Leffell DJ, Gerrard B, Goldstein AM, Dean M, Toftgard R, Chenevix-Trench G, Wainwright B, Bale

AE (1996) Mutations of the human homolog of drosophila patched in the nevoid basal cell carcinoma syndrome. Cell 85:841–851
12. Gorlin RJ (1995) Nevoid basal cell carcinoma syndrome. Dermatol Clin 13:113–125
13. Sun LS, Li XF, Li TJ (2008) PTCH1 and SMO gene alterations in keratocystic odontogenic tumors. J Dent Res 87:575–579
14. Gailani MR, Stahle-Backdahl M, Leffell DJ, Glynn M, Zaphiropoulos PG, Pressman C, Unden AB, Dean M, Brash DE, Bale AE, Toftgard R (1996) The role of the human homologue of Drosophila patched in sporadic basal cell carcinomas [see comments]. Nat Genet 14:78–81
15. Epstein EH (2008) Basal cell carcinomas: attack of the hedgehog. Nat Rev Cancer 8:743–754
16. Raffel C, Jenkins RB, Frederick L, Hebrink D, Alderete B, Fults DW, James CD (1997) Sporadic medulloblastomas contain PTCH mutations. Cancer Res 57:842–845
17. Reifenberger J, Wolter M, Weber RG, Megahed M, Ruzicka T, Lichter P, Reifenberger G (1998) Missense mutations in SMOH in sporadic basal cell carcinomas of the skin and primitive neuroectodermal tumors of the central nervous system. Cancer Res 58:1798–1803
18. Xie J, Murone M, Luoh SM, Ryan A, Gu Q, Zhang C, Bonifas JM, Lam CW, Hynes M, Goddard A, Rosenthal A, Epstein EH Jr, de Sauvage FJ (1998) Activating smoothened mutations in sporadic basal-cell carcinoma. Nature 391:90–92
19. Lam CW, Xie J, To KF, Ng HK, Lee KC, Yuen NW, Lim PL, Chan LY, Tong SF, McCormick F (1999) A frequent activated smoothened mutation in sporadic basal cell carcinomas. Oncogene 18:833–836
20. Brastianos PK, Horowitz PM, Santagata S, Jones RT, McKenna A, Getz G, Ligon KL, Palescandolo E, Van Hummelen P, Ducar MD, Raza A, Sunkavalli A, Macconaill LE, Stemmer-Rachamimov AO, Louis DN, Hahn WC, Dunn IF, Beroukhim R (2013) Genomic sequencing of meningiomas identifies oncogenic SMO and AKT1 mutations. Nat Genet 45:285–289
21. Clark VE, Erson-Omay EZ, Serin A, Yin J, Cotney J, Ozduman K, Avsar T, Li J, Murray PB, Henegariu O, Yilmaz S, Gunel JM, Carrion-Grant G, Yilmaz B, Grady C, Tanrikulu B, Bakircioglu M, Kaymakcalan H, Caglayan AO, Sencar L, Ceyhun E, Atik AF, Bayri Y, Bai H, Kolb LE, Hebert RM, Omay SB, Mishra-Gorur K, Choi M, Overton JD, Holland EC, Mane S, State MW, Bilguvar K, Baehring JM, Gutin PH, Piepmeier JM, Vortmeyer A, Brennan CW, Pamir MN, Kilic T, Lifton RP, Noonan JP, Yasuno K, Gunel M (2013) Genomic analysis of non-NF2 meningiomas reveals mutations in TRAF7, KLF4, AKT1, and SMO. Science 339:1077–1080
22. Mao J, Ligon KL, Rakhlin EY, Thayer SP, Bronson RT, Rowitch D, McMahon AP (2006) A novel somatic mouse model to survey tumorigenic potential applied to the hedgehog pathway. Cancer Res 66:10171–10178
23. Brugieres L, Remenieras A, Pierron G, Varlet P, Forget S, Byrde V, Bombled J, Puget S, Caron O, Dufour C, Delattre O, Bressac-de Paillerets B, Grill J (2012) High frequency of germline SUFU mutations in children with desmoplastic/nodular medulloblastoma younger than 3 years of age. J Clin Oncol 30:2087–2093
24. Taylor MD, Liu L, Raffel C, Hui CC, Mainprize TG, Zhang X, Agatep R, Chiappa S, Gao L, Lowrance A, Hao A, Goldstein AM, Stavrou T, Scherer SW, Dura WT, Wainwright B, Squire JA, Rutka JT, Hogg D (2002) Mutations in SUFU predispose to medulloblastoma. Nat Genet 31:306–310
25. Parsons DW, Li M, Zhang X, Jones S, Leary RJ, Lin JC, Boca SM, Carter H, Samayoa J, Bettegowda C, Gallia GL, Jallo GI, Binder ZA, Nikolsky Y, Hartigan J, Smith DR, Gerhard DS, Fults DW, VandenBerg S, Berger MS, Marie SK, Shinjo SM, Clara C, Phillips PC, Minturn JE, Biegel JA, Judkins AR, Resnick AC, Storm PB, Curran T, He Y, Rasheed BA, Friedman HS, Keir ST, McLendon R, Northcott PA, Taylor MD, Burger PC, Riggins GJ, Karchin R, Parmigiani G, Bigner DD, Yan H, Papadopoulos N, Vogelstein B, Kinzler KW, Velculescu VE (2011) The genetic landscape of the childhood cancer medulloblastoma. Science 331:435–439

26. Jones DT, Jager N, Kool M, Zichner T, Hutter B, Sultan M, Cho YJ, Pugh TJ, Hovestadt V, Stutz AM, Rausch T, Warnatz HJ, Ryzhova M, Bender S, Sturm D, Pleier S, Cin H, Pfaff E, Sieber L, Wittmann A, Remke M, Witt H, Hutter S, Tzaridis T, Weischenfeldt J, Raeder B, Avci M, Amstislavskiy V, Zapatka M, Weber UD, Wang Q, Lasitschka B, Bartholomae CC, Schmidt M, von Kalle C, Ast V, Lawerenz C, Eils J, Kabbe R, Benes V, van Sluis P, Koster J, Volckmann R, Shih D, Betts MJ, Russell RB, Coco S, Tonini GP, Schuller U, Hans V, Graf N, Kim YJ, Monoranu C, Roggendorf W, Unterberg A, Herold-Mende C, Milde T, Kulozik AE, von Deimling A, Witt O, Maass E, Rossler J, Ebinger M, Schuhmann MU, Fruhwald MC, Hasselblatt M, Jabado N, Rutkowski S, von Bueren AO, Williamson D, Clifford SC, McCabe MG, Collins VP, Wolf S, Wiemann S, Lehrach H, Brors B, Scheurlen W, Felsberg J, Reifenberger G, Northcott PA, Taylor MD, Meyerson M, Pomeroy SL, Yaspo ML, Korbel JO, Korshunov A, Eils R, Pfister SM, Lichter P (2012) Dissecting the genomic complexity underlying medulloblastoma. Nature 488:100–105
27. Pugh TJ, Weeraratne SD, Archer TC, Pomeranz Krummel DA, Auclair D, Bochicchio J, Carneiro MO, Carter SL, Cibulskis K, Erlich RL, Greulich H, Lawrence MS, Lennon NJ, McKenna A, Meldrim J, Ramos AH, Ross MG, Russ C, Shefler E, Sivachenko A, Sogoloff B, Stojanov P, Tamayo P, Mesirov JP, Amani V, Teider N, Sengupta S, Francois JP, Northcott PA, Taylor MD, Yu F, Crabtree GR, Kautzman AG, Gabriel SB, Getz G, Jager N, Jones DT, Lichter P, Pfister SM, Roberts TM, Meyerson M, Pomeroy SL, Cho YJ (2012) Medulloblastoma exome sequencing uncovers subtype-specific somatic mutations. Nature 488:106–110
28. Peukert S, Miller-Moslin K (2010) Small-molecule inhibitors of the hedgehog signaling pathway as cancer therapeutics. ChemMedChem 5:500–512
29. Han YG, Kim HJ, Dlugosz AA, Ellison DW, Gilbertson RJ, Alvarez-Buylla A (2009) Dual and opposing roles of primary cilia in medulloblastoma development. Nat Med 15:1062–1065
30. Ng JM, Curran T (2011) The hedgehog's tale: developing strategies for targeting cancer. Nat Rev Cancer 11:493–501
31. Yang L, Xie G, Fan Q, Xie J (2010) Activation of the hedgehog-signaling pathway in human cancer and the clinical implications. Oncogene 29:469–481
32. Berman DM, Karhadkar SS, Hallahan AR, Pritchard JI, Eberhart CG, Watkins DN, Chen JK, Cooper MK, Taipale J, Olson JM, Beachy PA (2002) Medulloblastoma growth inhibition by hedgehog pathway blockade. Science 297:1559–1561
33. Berman DM, Karhadkar SS, Maitra A, Montes De Oca R, Gerstenblith MR, Briggs K, Parker AR, Shimada Y, Eshleman JR, Watkins DN, Beachy PA (2003) Widespread requirement for hedgehog ligand stimulation in growth of digestive tract tumours. Nature 425:846–851
34. Watkins DN, Berman DM, Burkholder SG, Wang B, Beachy PA, Baylin SB (2003) Hedgehog signalling within airway epithelial progenitors and in small-cell lung cancer. Nature 422:313–317
35. Karhadkar SS, Bova GS, Abdallah N, Dhara S, Gardner D, Maitra A, Isaacs JT, Berman DM, Beachy PA (2004) Hedgehog signalling in prostate regeneration, neoplasia and metastasis. Nature 431:707–712
36. Weierstall U, James D, Wang C, White TA, Wang D, Liu W, Spence JC, Bruce Doak R, Nelson G, Fromme P, Fromme R, Grotjohann I, Kupitz C, Zatsepin NA, Liu H, Basu S, Wacker D, Han GW, Katritch V, Boutet S, Messerschmidt M, Williams GJ, Koglin JE, Marvin Seibert M, Klinker M, Gati C, Shoeman RL, Barty A, Chapman HN, Kirian RA, Beyerlein KR, Stevens RC, Li D, Shah ST, Howe N, Caffrey M, Cherezov V (2014) Lipidic cubic phase injector facilitates membrane protein serial femtosecond crystallography. Nat Commun 5:3309–3318
37. Chen JK, Taipale J, Cooper MK, Beachy PA (2002) Inhibition of hedgehog signaling by direct binding of cyclopamine to smoothened. Genes Dev 16:2743–2748
38. Jimeno A, Weiss GJ, Miller WH Jr, Gettinger S, Eigl BJ, Chang AL, Dunbar J, Devens S, Faia K, Skliris G, Kutok J, Lewis KD, Tibes R, Sharfman WH, Ross RW, Rudin CM (2013) Phase I study of the hedgehog pathway inhibitor IPI-926 in adult patients with solid tumors. Clin Cancer Res 19:2766–2774

39. Lee MJ, Hatton BA, Villavicencio EH, Khanna PC, Friedman SD, Ditzler S, Pullar B, Robison K, White KF, Tunkey C, LeBlanc M, Randolph-Habecker J, Knoblaugh SE, Hansen S, Richards A, Wainwright BJ, McGovern K, Olson JM (2012) Hedgehog pathway inhibitor saridegib (IPI-926) increases lifespan in a mouse medulloblastoma model. Proc Natl Acad Sci U S A 109:7859–7864
40. Low JA, de Sauvage FJ (2010) Clinical experience with hedgehog pathway inhibitors. J Clin Oncol 28:5321–5326
41. Von Hoff DD, LoRusso PM, Rudin CM, Reddy JC, Yauch RL, Tibes R, Weiss GJ, Borad MJ, Hann CL, Brahmer JR, Mackey HM, Lum BL, Darbonne WC, Marsters JC Jr, de Sauvage FJ, Low JA (2009) Inhibition of the hedgehog pathway in advanced basal-cell carcinoma. N Engl J Med 361:1164–1172
42. LoRusso PM, Rudin CM, Reddy JC, Tibes R, Weiss GJ, Borad MJ, Hann CL, Brahmer JR, Chang I, Darbonne WC, Graham RA, Zerivitz KL, Low JA, Von Hoff DD (2011) Phase I trial of hedgehog pathway inhibitor vismodegib (GDC-0449) in patients with refractory, locally advanced or metastatic solid tumors. Clin Cancer Res 17:2502–2511
43. Sekulic A, Migden MR, Oro AE, Dirix L, Lewis KD, Hainsworth JD, Solomon JA, Yoo S, Arron ST, Friedlander PA, Marmur E, Rudin CM, Chang AL, Low JA, Mackey HM, Yauch RL, Graham RA, Reddy JC, Hauschild A (2012) Efficacy and safety of vismodegib in advanced basal-cell carcinoma. N Engl J Med 366:2171–2179
44. Tang JY, Mackay-Wiggan JM, Aszterbaum M, Yauch RL, Lindgren J, Chang K, Coppola C, Chanana AM, Marji J, Bickers DR, Epstein EH Jr (2012) Inhibiting the hedgehog pathway in patients with the basal-cell nevus syndrome. N Engl J Med 366:2180–2188
45. Rudin CM, Hann CL, Laterra J, Yauch RL, Callahan CA, Fu L, Holcomb T, Stinson J, Gould SE, Coleman B, LoRusso PM, Von Hoff DD, de Sauvage FJ, Low JA (2009) Treatment of medulloblastoma with hedgehog pathway inhibitor GDC-0449. N Engl J Med 361:1173–1178
46. Yauch RL, Dijkgraaf GJ, Alicke B, Januario T, Ahn CP, Holcomb T, Pujara K, Stinson J, Callahan CA, Tang T, Bazan JF, Kan Z, Seshagiri S, Hann CL, Gould SE, Low JA, Rudin CM, de Sauvage FJ (2009) Smoothened mutation confers resistance to a hedgehog pathway inhibitor in medulloblastoma. Science 326:572–574
47. Atwood SX, Chang AL, Oro AE (2012) Hedgehog pathway inhibition and the race against tumor evolution. J Cell Biol 199:193–197
48. Chang AL, Oro AE (2012) Initial assessment of tumor regrowth after vismodegib in advanced basal cell carcinoma. Arch Dermatol 148:1324–1325
49. Brinkhuizen T, Reinders MG, van Geel M, Hendriksen AJ, Paulussen AD, Winnepenninckx VJ, Keymeulen KB, Soetekouw PM, van Steensel MA, Mosterd K (2014) Acquired resistance to the hedgehog pathway inhibitor vismodegib due to smoothened mutations in treatment of locally advanced basal cell carcinoma. J Am Acad Dermatol 71:1005–1008
50. Buonamici S, Williams J, Morrissey M, Wang A, Guo R, Vattay A, Hsiao K, Yuan J, Green J, Ospina B, Yu Q, Ostrom L, Fordjour P, Anderson DL, Monahan JE, Kelleher JF, Peukert S, Pan S, Wu X, Maira SM, Garcia-Echeverria C, Briggs KJ, Watkins DN, Yao YM, Lengauer C, Warmuth M, Sellers WR, Dorsch M (2010) Interfering with resistance to smoothened antagonists by inhibition of the PI3K pathway in medulloblastoma. Sci Transl Med 2:51–70
51. McMillan R, Matsui W (2012) Molecular pathways: the hedgehog signaling pathway in cancer. Clin Cancer Res 18:4883–4888
52. Lin TL, Matsui W (2012) Hedgehog pathway as a drug target: smoothened inhibitors in development. Onco Targets Ther 5:47–58
53. Hadden MK (2013) Hedgehog pathway inhibitors: a patent review (2009–present). Expert Opin Ther Pat 23:345–361
54. Kim J, Tang JY, Gong R, Kim J, Lee JJ, Clemons KV, Chong CR, Chang KS, Fereshteh M, Gardner D, Reya T, Liu JO, Epstein EH, Stevens DA, Beachy PA (2010) Itraconazole, a commonly used antifungal that inhibits hedgehog pathway activity and cancer growth. Cancer Cell 17:388–399

55. Hoch L, Faure H, Roudaut H, Schoenfelder A, Mann A, Girard N, Bihannic L, Ayrault O, Petricci E, Taddei M, Rognan D, Ruat M (2015) MRT-92 inhibits hedgehog signaling by blocking overlapping binding sites in the transmembrane domain of the smoothened receptor. FASEB J
56. Kim DJ, Kim J, Spaunhurst K, Montoya J, Khodosh R, Chandra K, Fu T, Gilliam A, Molgo M, Beachy PA, Tang JY (2014) Open-label, exploratory phase II trial of oral itraconazole for the treatment of basal cell carcinoma. J Clin Oncol 32:745–751
57. Gorojankina T, Hoch L, Faure H, Roudaut H, Traiffort E, Schoenfelder A, Girard N, Mann A, Manetti F, Solinas A, Petricci E, Taddei M, Ruat M (2013) Discovery, molecular and pharmacological characterization of GSA-10, a novel small-molecule positive modulator of smoothened. Mol Pharmacol 83:1020–1029
58. Teperino R, Amann S, Bayer M, McGee SL, Loipetzberger A, Connor T, Jaeger C, Kammerer B, Winter L, Wiche G, Dalgaard K, Selvaraj M, Gaster M, Lee-Young RS, Febbraio MA, Knauf C, Cani PD, Aberger F, Penninger JM, Pospisilik JA, Esterbauer H (2012) Hedgehog partial agonism drives Warburg-like metabolism in muscle and brown fat. Cell 151:414–426
59. Robbins DJ, Fei DL, Riobo NA (2012) The hedgehog signal transduction network. Sci Signal 5:re6
60. Kenakin T, Christopoulos A (2013) Signalling bias in new drug discovery: detection, quantification and therapeutic impact. Nat Rev Drug Discov 12:205–216
61. Barzi M, Kostrz D, Menendez A, Pons S (2011) Sonic hedgehog-induced proliferation requires specific Galpha inhibitory proteins. J Biol Chem 286:8067–8074
62. Shen F, Cheng L, Douglas AE, Riobo NA, Manning DR (2013) Smoothened is a fully competent activator of the heterotrimeric G protein G(i). Mol Pharmacol 83:691–697

Top Med Chem (2015) 16: 13–42
DOI: 10.1007/7355_2014_63

Published online: 25 October 2014

Canonical and Non-Canonical Hedgehog Signaling Pathways: Role of G Proteins

Natalia A. Riobo

Abstract The Hedgehog (Hh) signaling pathway has received a great deal of attention in the past decade due to its involvement in cancer, angiogenesis, and fibrosis. Several inhibitors of the pathway were developed which target the 7-transmembrane protein Smoothened (Smo), core component of the canonical pathway that controls Gli-dependent transcriptional activity. However, recent studies revealed that the Hh pathway has other transcription-independent functions, collectively known as "non-canonical signaling." This review describes the role and function of each Hh pathway component in canonical and non-canonical signaling, with emphasis on the role of Smo as a GPCR that selectively activates heterotrimeric G_i proteins.

Keywords Gli, Hedgehog, Heterotrimeric G proteins, Non-canonical, Smoothened

Contents

N.A. Riobo (✉)
Department of Biochemistry and Molecular Biology and Kimmel Cancer Center, Sidney Kimmel Medical College, Thomas Jefferson University, Philadelphia, PA, USA
e-mail: natalia.riobo@jefferson.edu

Abbreviations

12TM	12 transmembrane
20(S)OHC	20(S)-Hydroxycholesterol
25OHC	25-hydroxycholesterol
5HT	5-hydroxytriptamine
7TM	7 transmembrane
AMP	Adenosine monophosphate
cAMP	3′-5′-cyclic adenosine monophosphate
CCK	Cholecystokinin
ChIP	Chromatin immunoprecipitation
CoA	Coenzyme A
cyclopamine	(2′*R*,3*S*,3′*R*,3′a*S*,6′*S*,6a*S*,6b*S*,7′a*R*,11a*S*,11b*R*)-1,2,3,3′a,4,4′,5′,6,6′,6a,6b,7,7′,7′a,8,11,11a,11b-Octadecahydro-3′,6′,10,11b-tetramethylspiro[9*H*-benzo[*a*]fluorene-9,2′(3′*H*)-furo[3,2-*b*]pyridin]-3-ol
GDC-0449	(2-chloro-N-[4-chloro-3-pyridin-2-yl-phenyl]-4-methane-sulfonyl benzamide)
GSA-10	4-[[(1-Hexyl-1,2-dihydro-2-oxo-3-quinolinyl)carbonyl]amino] benzoic acid propyl ester
GTP	Guanosine-5′-triphosphate
GTPγ[S]	Guanosine 5′-O-[γ-thio]triphosphate
H89	N-[2-(p-Bromocinnamylamino)ethyl]-5-isoquinolinesulfonamide
LDE-225	N-(6-((2S,6R)-2,6-dimethylmorpholino)pyridin-3-yl)-2-methyl-4′-(trifluoromethoxy)biphenyl-3-carboxamide
M25	*N*-[[1-(2-Methoxyphenyl)-1*H*-indazol-5-yl]methyl]-2-propylpentanamide
NES	Nuclear export signal

PI4P	Phosphatidylinositol 4-phosphate
SAG	N-Methyl-N′-(3-pyridinylbenzyl)-N′-(3-chlorobenzo[b]thiophene-2-carbonyl)-1,4-diaminocyclohexane
SANT-1	(4-Benzyl-piperazin-1-yl)-(3,5-dimethyl-1-phenyl-1H-pyrazol-4-ylmethylene)-amine
siRNA	Small interference RNA
XL-139	N-(2-methyl-5-((methylamino)methyl)phenyl)-4-((4-phenylquinazolin-2-yl)amino)benzamide

1 Introduction

Hedgehog signaling is initiated by binding of any of three Hedgehog (Hh) isoforms to their receptors Patched-1 or Patched-2. Binding of the ligand results in inhibition of a repressive activity of Patched proteins, still molecularly undefined, which results in a cellular response. The nature of that response is variable in a context- and cell type-specific manner. An artificial classification of the molecular entities and responses activated downstream of Patched-1 distinguishes a "canonical" pathway from less well-defined "non-canonical" responses (Fig. 1) [1–4]. The canonical Hh pathway, initially characterized in the fruit fly, has crucial roles in vertebrate developmental biology. Canonical signaling involves de-repression of the 7TM protein Smoothened (Smo) following Hh binding to Patched-1, and ultimately results in activation of the Gli family of transcription factors and transcription-mediated cellular responses such as cell fate specification and proliferation. It takes place at the primary cilium and requires the participation of co-receptor proteins for the formation of the signaling ligand/receptor complex (Fig. 2). Non-canonical Hh signaling, instead, leads to very rapid, Gli transcription-independent responses of diverse types. Type I signaling results in inhibition of a pro-apoptotic activity of Patched-1, upstream of Smo activation, while type II signaling occurs downstream of Smo and requires coupling of Smo to heterotrimeric G inhibitory (G_i) proteins. The latter have been shown to activate Rho GTPases leading to actin cytoskeleton changes and to induce calcium uptake via L-type calcium channels (LTCC), modulating metabolic responses.

The novel role of Patched-1, and particularly of Smo, as regulators of a myriad of clinically important cellular functions that are independent of induction of Gli1, a hallmark of canonical Hh signaling that is used as a measure of efficacy in drug development, advocates a need for reassessment of the classification of past and novel small molecule regulators of the Hh pathway.

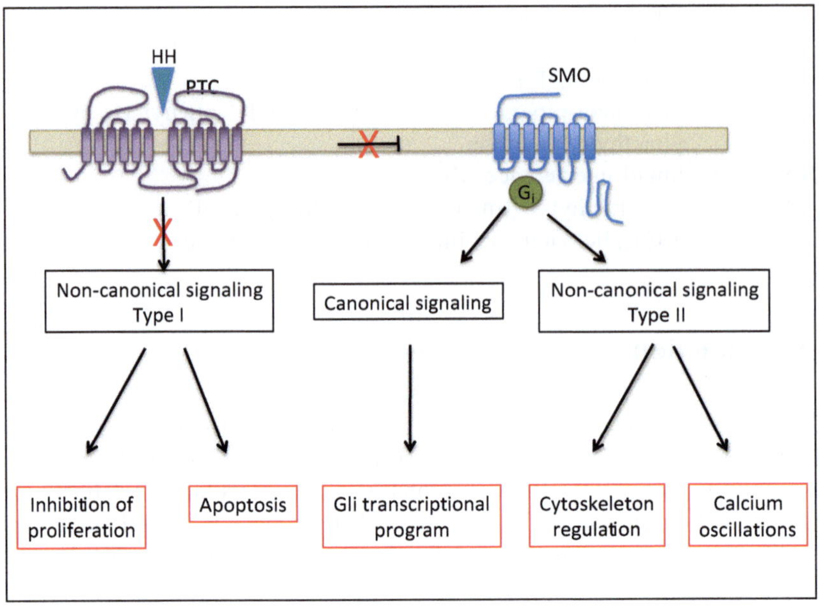

Fig. 1 Classification of the Hedgehog signaling network. *HH* any Hedgehog ligand, *PTC* Patched, *SMO* Smoothened; *Gi* any Gi protein

2 Hedgehog Proteins

2.1 Structural Features

Three Hh isoforms exist in mammals: Sonic Hh (Shh), Indian Hh (Ihh), and Desert Hh (Dhh). While they share a high degree of homology, the most similar to Drosophila Hh is Dhh. All genes encode precursor proteins of ~45 kDa, which are processed into signaling active N-terminal fragments of 19–20 kDa. The active Shh-N polypeptide has been crystalized at 1.7 Å resolution and the structure revealed that Shh structure is a core $\alpha+\beta$ sandwich with two α-helixes and six mixed β-sheets connected by several loops [5]. Remarkably, the crystal revealed a protein fold consisting of a metalloprotease-like tetrahedrally coordinated Zn^{2+} ion, with homology to Zn^{2+} hydrolases like carboxypeptidase A but which was later determined to be a pseudo active site [5, 6]. The C-terminal portion of the Drosophila Hh precursor (Hh-C), which encodes the autoprocessing domain, has also been crystalized showing an all β-sheet disc-like structure composed of two homologous subdomains related by a pseudo twofold axis of symmetry, suggesting that they could have arisen by tandem duplication of a primordial gene [7]. While there is minimal homology to other self-processing proteins, His 329 in Drosophila Hh-C is conserved in all Hh isoforms and in the active site of all other inteins, and is required for thioester formation activity. Another residue, Thr 326 is also conserved and required for efficient catalysis.

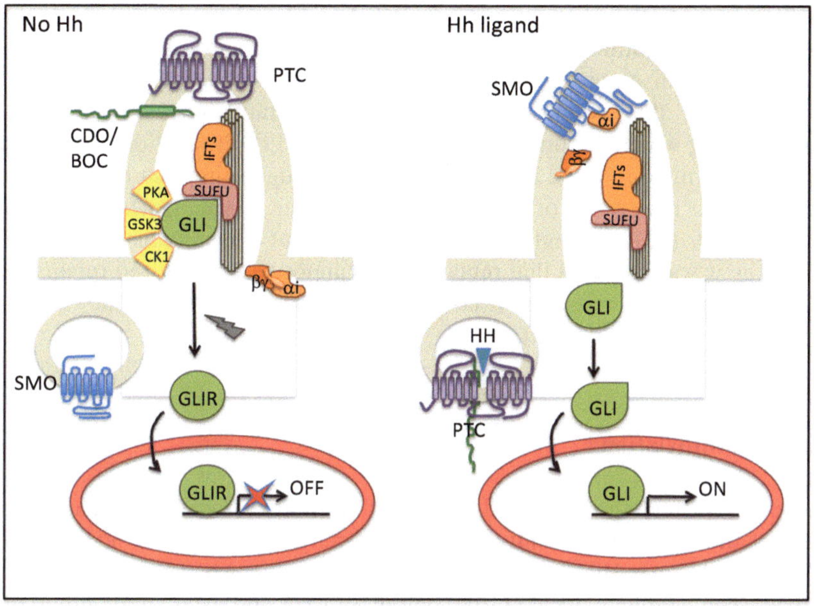

Fig. 2 Vertebrate canonical Hedgehog signaling. *Left*: in the absence of a Hh ligand Patched1 (PTC) accumulates in the primary cilium and represses Smoothened (SMO) activity. Full length GLI2 and GLI3 (GLI) are processed in a cilium-dependent manner into repressors (GLIR). *Right*: Hh binding to PTC and the co-receptors (CDO/BOC/GAS1) causes its endocytosis and translocation of SMO to the cilium in an active conformation that couples to G_i proteins (αiβγ). SMO activation prevents GLI processing and induces its full activation, subsequently activating Hh-target genes

2.2 *Synthesis and Lipid Modifications*

2.2.1 Cleavage and Cholesterol Modification

Hh proteins undergo extensive posttranslational processing and modifications before they are secreted from the producing cells. The 45 kDa precursor is synthesized in the ER and the signal peptide is rapidly cleaved. The newly generated N-terminus starts at former Cys 24 (position refers to that of Shh). The protein then suffers an ancient and rare self-processing modification, in which the intein-like domain located in the C-terminal region of Hh catalyzes the cleavage of the precursor at the Gly 197-Cys 198 peptide bond to generate the active Hh-N fragments. The autoprocessing reaction proceeds in two steps. First, the thiol group of the conserved Cys 198 acts as a nucleophile to attack the carbonyl group of the preceding Gly residue and forms a thioester linkage in place of the peptide bond. Second, a nucleophilic attack from the 3β hydroxyl group of a cholesterol molecule breaks apart the thioester bond. This reaction results in the release of the C-terminal domain of Hh (Hh-C) and formation of a covalent ester adduct between cholesterol and the C-terminus of Hh-N. Many nucleophiles, including dithiothreitol, glutathione, and hydroxylamine, can substitute for cholesterol and stimulate Hh autoprocessing in vitro, confirming the mechanism [8]. The importance of this reaction was revealed by mutations that impair Shh autoprocessing which disrupt embryonic patterning [9]. Production of cholesterol-free Shh-N in mice resulted in patterning defects, suggesting that the hydrophobic modification is necessary for proper formation of the morphogenetic gradient [10, 11]. However, the full-length, unprocessed form of Shh retains some localized activity, as it was reported that mutants that cannot be proteolyzed retain activity when the producing cells are in contact with reporter cells, but they cannot be released into the medium [12].

2.2.2 Palmitoylation

Hh precursors undergo a second important posttranslational modification in the lumen of the ER, independently of the processing by cholesterol addition to Gly 197 described above. Hh is palmitoylated at the N-terminus by formation of an amide bond between palmitate donated from palmitoyl-CoA and the α amino group of Cys 24 (in human Shh, position 25 in mouse) [13]. This modification is irreversible, in contrast to the more common S-palmitoylation found in internal residues. Two possible mechanisms of palmitoylation have been proposed. In the first mechanism, palmitoylation occurs via a thioester linkage to Cys 24, followed by a rearrangement of the thioester intermediate to an amide linkage via an intramolecular S-to-N shift, producing an amide or N-linked palmitate [13, 14]. In the second mechanism, palmitate will be directly attached to the N-terminal amide via amide linkage, analogous to *N*-myristoylation [15]. Shh modification is catalyzed by Hh acyltransferase (Hhat). Hhat is a ~50 kDa protein belonging to the MBOAT family of acyltransferases which resides in the ER/Golgi, and predicted to contain at least 8 TM segments [16]. The second mechanistic model is supported by the findings that N-terminally blocked Shh is not substrate for Hhat and that thioester linked palmitoylated intermediates of Shh have not been detected [17]. Mutation of the acceptor Cys 24 in Shh results in reduced signaling capacity [18]. The hydrophobic nature of palmitate appears to directly contribute to its ability to enhance Shh potency, as replacement of palmitate by an Ile-Ile dipeptide confers increased activity compared to non-palmitoylated Shh [13]. Indeed, this modification has been exploited to produce active recombinant Shh, Ihh, and Dhh [19].

2.3 *Release of Hedgehog from Producing Cells and Modes of Short and Long Distance Signaling*

Since Hh ligands are covalently modified by cholesterol and palmitate, they primarily localize to the plasma membrane of expressing cells. Nonetheless, they can act at a distance from the source. Several mechanisms for long-distance Hh signaling have been proposed, including diffusion of free solubilized Hhs, vesicles containing Hhs (exosomes), or long cell extensions (cytonemes).

2.3.1 Role of Scube/You, Dispatched-A and Dispatched-B

Release of free Hh molecules into the extracellular milieu requires the action of a group of proteins that serve to detach the ligands from the cell surface of producing cells. The first one to be discovered was Dispatched (Disp) in Drosophila, highly homologous to the Hh receptor Patched. Two redundant homologs in mammals, called DispA and DispB, were shown to be required for the release of fully lipid-modified Shh. Deficiency of DispA and B in mice phenocopies the loss of long-range signaling of multiple Hh family members [20–22]. Disp proteins act in concert with Scube2 (homolog of the zebrafish *you* mutant) to synergistically release dual lipid-modified Hh proteins to the medium [23]. Scube2 is also secreted into the medium and can promote Hh release in a non-cell autonomous manner [23]. However, the exact mechanism of action of Dispatched proteins and Scube is still unknown.

2.3.2 Multimerization

Once they are released from the producing cells, Hh proteins can be isolated from tissue culture cell supernatants in high-molecular-weight multimeric complexes of sizes ranging between 158 and 4,000 kDa (from 6 to at least 160 times the monomer size) [24]. It is likely that the lipids are nucleated in the core and the hydrophilic polypeptide surfaces are exposed to the aqueous millieu. While the multimers are signaling-competent, monomeric Hhs do not signal very efficiently. Formation of the high molecular-weight complexes depends on both palmitoylation of the N-terminus and cholesterol modification of the C-terminus [18, 24, 25].

2.3.3 Exosomes

In addition to free monomers/multimers, Hh proteins can spread through space associated with small vesicles derived from budding of the plasma membrane of producing cells, called microvesicles or exosomes. It was shown that stimulated T cells shed microvesicles containing active Shh, detected both by western blot and by activity assays [26]. These vesicle carriers stimulate neovascularization in a Hh-dependent manner [27, 28]. Hepatic stellate cells have also been shown to shed microvesicles and exosomes containing Hh ligands under certain conditions, which stimulate angiogenesis in the liver [29]. Microvesicles containing Shh have also been isolated from the circulation, suggesting that they might mediate very long-range signaling [27].

2.3.4 Cytonemes

Recent elegant studies in Drosophila and chick embryos have revealed that Hh can signal at a distance despite remaining associated with the plasma membrane of the producing cell [30–32]. Cytonemes are a specialized class of actin-based filopodia that span several cell diameters and transport components of the Hh pathway. In both organisms there are strong evidences of highly stabilized interactions between cytonemes containing Hh ligand and those containing Patched/Patched-1 and co-receptors, suggesting that long-range signaling may be mediated through direct receptor–ligand interactions between cell membranes at a distance. In the chick embryo limb bud, Shh-producing and target cells can extend filopodia as long as 150 μm in length [32]. Moreover, formation of the Hh gradient correlates with cytoneme formation in space and time, and mutations that affect cytoneme formation reduce Hh gradient length in the fruit fly [31]. The evidence supports the notion that cytoneme-mediated Hh transport is the mechanistic basis for Hh gradient formation.

3 The Hedgehog Receptors Patched-1 and Patched-2

3.1 Predicted Topology and Domains

The Hh receptors Patched-1 and Patched-2 have a predicted 12TM topology with intracellular N- and C-terminal domains, six extracellular loops, and five intracellular loops, of which the central intracellular loop is the longest. They have homology with the resistance-nodulation-cell division (RND) family efflux pumps responsible for much of the multidrug resistance of Gram-negative bacteria [33]. Not only is the overall topology of Patched-1 and -2 similar to that of bacterial RND permeases, but also they have conserved aspartic acid residues in the TM4, which are thought to be important for proton translocation in the bacterial permeases. Patched isoforms are also characterized by the presence of a sterol-sensing domain (SSD). The SSD is a region of five consecutive membrane-spanning domains that is conserved across a range of proteins with roles in sterol and vesicle trafficking [33]. The role of this domain appears to be different in Drosophila and in mammals. SSD mutations in TM3 and TM6 of Drosophila Patched were shown to behave as dominant-negative alleles [34]. In contrast to the clear dominant-negative effect observed in Drosophila, the corresponding mutation in mouse *Ptc1* did not significantly alter Patched-1 function in an in vitro assay [35]. However, a single mutant isolated from a patient with Nevus Basal Cell Carcinoma Syndrome (NBCCS) harboring the single substitution of Gly509Val in the TM4 of Patched-1 has dominant negative activity when assayed in the Drosophila wing disc [36]. These observations suggest that the SSD of Patched is involved in Smo regulation in vertebrates but that despite the sequence homology the role of specific residues is not the same that in Drosophila.

The intracellular C-terminal domain (CTD) of Patched-1 encodes additional functions. Deletion of the CTD in Drosophila Patched affects localization and results in plasma membrane accumulation [37]. In mammals, the CTD of Patched-1 serves as a site of pro-caspase 9 recruitment and initiates apoptosis when overexpressed unless it is bound to a Hh ligand [38, 39]. The CTD of Patched-1 contains a PPXY sequence, a putative target site for Nedd4 ubiquitin ligases, and its deletion renders the receptor more stable and at higher levels in the plasma membrane [37]. CTD deletions are more stable than the wild-type protein [37]. Several mutations in the CTD of human Patched-1 have been identified in human cancers [40].

3.2 Catalytic Activity of Patched-1 Isoforms

The finding that Patched can repress Smo activation in substoichiometric concentrations, estimated in 1 to 250, and the fact that Patched and Smo traffic in opposite directions in the cells strongly suggests that Patched catalytically inhibits Smo [41, 42].

3.2.1 Regulation of Phosphatidylinositol 4 III Kinases

A candidate regulator of Smo activation is the intracellular level of phosphatidylinositol-4-phosphate (PI4P) [43]. The lipid kinase that generates PI4P from PI is Stt4 in Drosophila and a group of PI4 II and III kinases in vertebrates. It was shown that Drosophila Patched reduces the level of PI4P in a Stt4-dependent manner, suggesting that Patched directly or indirectly inhibits Stt4 in the absence of Hh [43]. Moreover, siRNA knockdown experiments in NIH3T3 fibroblasts showed that depletion of PI4 IIIα and β reduces activation of a Gli-luciferase reporter by Shh, indicating that the mechanism is at least partly conserved [43].

3.2.2 Regulation of Vitamin D Metabolites

The SSD of Patched brought attention to the possibility that it regulates transport of sterol-like compounds towards the inside or outside of the cell. A different way of regulation of Smo by Patched-1, involving Patched-1-dependent secretion/extrusion of pro-vitamin D3 has been proposed [44]. However, that study proposes that Patched-1 can function in a non-cell autonomous way, which is in conflict with a vast number of studies in vivo and in vitro and with the notion that Patched-1 localizes to the primary cilium. Nonetheless, vitamin D3 is a potent inhibitor of the Hh pathway downstream of Patched-1 and has been used to reduce proliferation of Hh-dependent cancers [45–47]. Importantly, the structural requirements of vitamin D3 for Hh pathway inhibition are distinct than those required for activation of the vitamin D receptor (VDR), suggesting that vitamin D3 may target a component of the Hh pathway [48].

In addition, the notion of Patched-1 as a sterol transporter is gaining strength since an independent study showed accumulation of cholesterol in cells expressing Patched-1 [49]. Whether these sterols are the physiological means of Smo modulation or represent novel functions of Patched-1 is still unknown.

3.3 *Subcellular Trafficking of Patched-1*

3.3.1 Trafficking to the Primary Cilium

In the past decade a great deal of new information about Patched-1 trafficking was published. The starting point for studying the role of cilia in Hh signal transduction was the identification of mouse mutants in ciliary proteins with phenotypes resembling of defective canonical Hh signaling [50, 51]. It was later discovered that under resting conditions Patched-1 accumulates at the tip of primary cilia, and that it exits the cilium in the presence of Shh [51]. However, the larger fraction of Patched-1 remains in equilibrium between the plasma membrane and intracellular vesicles, as it is the case in Drosophila, in which cells do not contain primary cilia. It is possible that plasma membrane Patched-1 in vertebrates has a different function as a mediator of non-canonical Hh signaling.

3.3.2 Endocytosis

Patched-1 is actively endocytosed and its half-life is rather short, estimated to be around 40 min by metabolic labeling [52]. A recent report in Drosophila suggests that Patched is internalized from the plasma membrane and targeted for degradation by a mechanism involving ubiquitination of a C-terminal Lys residue by the E3 ligase Smurf [53]. However, since mammalian canonical Hh signaling requires internalization of Patched-1 from the primary cilium membrane, the relevance of this mechanism in mammals is still unknown.

4 The Co-Receptors Cdo, Boc, and Gas-1

Activation of the canonical Hh signaling pathway was affected in different degrees by mutations of three single transmembrane proteins named Cell adhesion molecule, Down-regulated by Oncogenes (CDO), Brother Of *CDO* (*BOC*), and Growth Arrest-Specific 1 (GAS1). Mouse models lacking one or more of them showed that they have a positive redundant function in Hh signal transduction [54, 55].

CDO and BOC belong to the immunoglobulin (Ig) superfamily of membrane proteins [56]. They are characterized by having a single transmembrane domain, an extracellular N-terminus containing Ig and fibronectin type III domains (FNIII) and a cytoplasmic tail [57]. CDO contains five Ig followed by three FNII domains, of which the third FNIII domain has been mapped as the Shh-binding region [54]. Binding of CDO (and probably of BOC) to Shh occurs in a calcium-dependent manner and is contingent on the integrity of a calcium-binding pocket in Shh [57]. Conversely, the CDO/BOC homolog in Drosophila, Interference Hh (Ihog), binds to Hh through the first FNIII domain and in a heparin-dependent manner [58].

GAS1 is a glycosylphosphatidylinositol (GPI)-anchored cell surface protein, upregulated during cell cycle arrest that has been shown to interact with Shh and to stimulate Hh signal transduction [59, 60]. Mutations in all three co-receptors have been linked to holoprosencephaly and also bradydactily [61–63], and as mentioned earlier, deficiency in the three co-receptors abolishes all canonical Hh signaling during embryonic development [55].

5 The Core Transducer Smoothened

5.1 Overall Structure

The overall structural features of Smo reveal a 7-transmembrane (7TM) topology that resembles that of G protein-coupled receptors (GPCRs). The GPCR superfamily is subdivided into seven families sharing homologous domains in the extracellular N-terminal region and intracellular C-terminal tail and having a similar type of ligand [64]. Smo belongs to the F subfamily of GPCRs, which also contains the Frizzled (Fz) Wnt receptors. Smo has a characteristic extracellular cysteine-rich domain (CRD) and a long intracellular C-tail. Smo exists as a homodimer making contacts through both the N-terminal and the C-terminal domains, such as reported for many GPCRs [65]. The long C-tail of Smo undergoes a conformational change upon activation when the strong positive charge of an Arg-rich stretch is neutralized by phosphorylation of a large number of intercalated Ser/Thr residues [65].

5.1.1 The 7-Transmembrane Core

Interestingly, the crystal structure of the 7TM bundle resembles in topology that of group A GPCRs, which use peptides for ligands, despite a mere 10 % sequence homology [66]. An eighth short intracellular α-helix that lies parallel to the lipid bilayer is also characteristic of group A GPCRs. The 7TM core of Smo contains the cyclopamine binding pocket, which also binds the inverse agonist SANT-1 and the small molecule agonist SAG [67].

5.1.2 The N-Terminal Domain

The CRD of Smo is a compact globular domain, highly homologous to that of Fz receptors, containing four α-helixes and 2 short β-barrel motifs. Smo also contains large extracellular loops that make contacts with the CRD linker region and with TM helix III through a network of disulfide bonds [67]. The CRD and the extracellular loops create a lid that partly occludes the cyclopamine-binding pocket [67]. The binding site of oxysterols, small agonists of Smo, which allosterically modulate the cyclopamine-binding pocket, has been mapped to the CRD [68, 69]. Since some endogenous oxysterols activate canonical Hh signaling, it had been proposed that they could be the endogenous ligands of Smo, regulated somehow by Patched [70–72]. However, a recent study shows that deletion of the oxysterol-binding domain of Smo or point mutations that abolish binding cannot suppress Smo activation by Shh or SAG in vivo, although they abrogate its modulation by exogenous oxysterols [67]. This conclusive result indicates that oxysterols are modulators of Smo activity but are not the missing link between Patched and Smo.

5.1.3 Features of the C-Terminal Domain

The long intracellular C-tail of Smo is required for phosphorylation by G protein-coupled receptor kinase 2 (GRK2) and casein kinase 1 (CK1), for β-arrestin binding, trafficking to the primary cilium, and for interaction with Costal2 and Fused in Drosophila [73–78]. GRK2 phosphorylates many GPCRs in the C-tail when they acquire the active conformation. GRK2 is recruited to the active GPCR by binding to Gβγ transiently dissociated from heterotrimeric (Gαβγ) G proteins, supporting the notion that Smo is a *bona fide* GPCR. In Drosophila Smo, mutagenesis studies identified two GRK2 and GRK5 phosphorylation sites: Ser741/Thr742 and Ser1013/Ser1015 [73]. Mutation of those sites reduces Smo dimerization and blocks high Hh signaling. CK1 also phosphorylates vertebrate and Drosophila Smo. In vertebrate Smo there are conserved CK1 and GRK2 phosphorylation sites, and elegant studies have shown that Ser592, Ser594, Thr597, Ser599, Ser774, and Ser777 are CK1 sites and Ser642 and Ser666 are phosphorylated not only by GRK2 in vivo but also by GRK5 in vitro [79]. Interestingly, there seems to be a graded increase in Smo C-tail phosphorylation by increasing concentrations of Shh, suggesting that differential phosphorylation might be related to decoding of the morphogenetic signal. Phosphorylation of the C-tail of Smo is a pre-requisite for interaction of Smo with β-arrestin and a required modification to allow translocation of active Smo into the primary cilium [74, 76, 79].

5.2 Endocytosis and Degradation

Signal termination at the level of Smo had not been intensely studied until recently. Smo is internalized upon phosphorylation and monoubiquitination in multiple residues [80]. Inhibition of CK1 and cAMP-dependent protein kinase (PKA) with CK7 and H89, respectively, abolished ubiquitination; however, at the concentration used (10 μM), H89 also inhibits GRK2, thus it is likely that phosphorylation of Smo by CK1 and GRK2 promotes the subsequent ubiquitination and endocytosis. Endocytosed Smo traffics through early endosomes and late endosomes and is finally degraded by both proteasomal and lysosomal pathways [81]. Activation of the Hh pathway reduces Smo ubiquitination and degradation by stimulation of the deubiquitinase USP8 [80].

5.3 Coupling to Heterotrimeric G Proteins

Despite Smo is still orphan of an endogenous, Hh-dependent small ligand, there are clear evidences that it activates heterotrimeric G proteins, like all GPCRs. Biochemically, Smo is able to stimulate GDP-GTP exchange in all members of the G inhibitory (G_i) family of G proteins (G_{i1}, G_{i2}, G_{i3}, G_o, and G_z) but not of any other G protein families in the absence of Patched or when activated by the small agonist purmorphamine [82]. Interestingly, the C-tail partly precludes interaction with G_i proteins, since deletion of the C-tail increases GTPγ[S^{35}] binding, suggesting that activation of Smo opens up its conformation and allows interaction with G proteins [82]. The strong selectivity of Smo towards the G_i family is comparable to the prototypical G_i-coupled receptor 5HT1aR, with equimolar potency for G_i activation, while inhibition of cAMP production by Smo is more potent than that of 5HT1aR [83]. Importantly, while Smo is a strong activator of G_i proteins, G_i activation alone is not sufficient for activation of the canonical Hh pathway downstream of Smo, and in some cells types is not even necessary [82, 84]. The use of a Pertussis toxin (PTX) has served to demonstrate that Gli activation in NIH3T3 cells, CH10T1/2 cells, and cardiomyocytes is largely sensitive to inhibition of G_i protein activation [82, 83, and C. Carbe et al., unpublished data]. PTX ADP-ribosylates the N-terminal Cys residue in the α subunit of G_i isoforms preventing their interaction with GPCRs. In other cell types, such as mouse embryonic fibroblasts (MEFs), Smo-G_i coupling is not essential for canonical Hh signaling, and the reason is still unclear [84]. In Drosophila S2 cells, depletion of the single G_i family member by siRNA prevents activation of Ci by Hh and deficiency of G_i in wing disc clones inhibits expression of the Hh-target gene *dpp* [85]. PTX has also been used with conflicting results in in vivo studies in zebrafish and avian embryos. In zebrafish, PTX phenocopies many aspects of Shh deficiency including cyclopia, lack of forebrain ventral specification, and an expansion of the sclerotome at the expense of adaxial fates in the posterior somites [86]. In the chick embryo neural tube electroporation model, PTX did not affect gross dorso-ventral patterning and thus it was concluded that G_i is not required for Hh-dependent patterning [87]. However, the nervous system expresses G_z, a member of the G_i family that couples to Smo but is insensitive to PTX because it lacks the Cys residue that is the substrate for ADP-ribosylation. Therefore, the results are inconclusive with respect to the role of G_i proteins in neural tube patterning by Shh. Proliferation of mammalian cerebellar granule neuron precursors in response to Shh, mediated by Gli transcription, also depends on G_i proteins as is inhibited by PTX and by siRNA-mediated depletion of $G\alpha_{i1}$, $G\alpha_{i2}$, or $G\alpha_{i3}$. In this model system, overexpressed $G\alpha_{i1}$ and $G\alpha_{i2}$ where detected in different regions of the primary cilium [88].

6 Suppressor of Fused

6.1 Structure and Features

Suppressor of Fused (Sufu) is a negative regulator of the Hh pathway and its function is essential for mammalian Hh signaling, as inactivation of *Sufu* in mice leads to ectopic activation of the Hh pathway and embryonic lethality [89–91]. Sufu is a 484-aminoacid protein with no significant homology to any other protein except for a conserved PEST domain [91, 93]. The crystal structure of Sufu revealed two separated N- and C-terminal halves linked by a disordered region that clamp around the Gli transcription factors through a conserved SYGHL motif in Gli [94, 95].

6.2 Regulation of Gli Stability and Nuclear Localization

In the absence of Hh, Sufu binds directly to Gli proteins and retains them in the cytoplasm to prevent pathway activation in both Drosophila and vertebrates [92, 96–100]. Sufu binds to the N-terminal ends of the Gli proteins via interaction with its C-terminal domain, while Sufu's N-terminal domain binds the C-tail of Gli1. While each domain is capable of binding to different Gli1 regions independently, interactions between Sufu and Gli1 at both sites are required for cytoplasmic tethering and repression of Gli1 [101]. In addition, the kinase Cdc2l1 that was singled out in a siRNA screen designed to identify kinases affecting Hh signaling, was found to relieve the inhibitory effect of Sufu on Gli-dependent transcription by binding to Sufu and modulating the Gli-Sufu interaction [102].

In addition to cytoplasmic retention of Glis, Sufu plays a key role in the stabilization of the activated forms of Gli2 and Gli3, but not of the processed repressors of transcription. In Sufu-null cells, the half-lives of Gli proteins are dramatically reduced. It was found that depletion of the Cul3 adaptor protein speckle-type POZ protein (SPOP) was sufficient to restore full length Gli2 and Gli3 levels in Sufu$^{-/-}$ cells. Consistent with this, Sufu regulates Gli2 and Gli3 protein levels by antagonizing the activity of SPOP [103, 104].

6.3 Regulation of Gli Activation at the Primary Cilium

Besides preventing nuclear translocation of the Gli transcription factors and promoting their phosphorylation by PKA and Glycogen Synthase Kinase-3β (GSK3β), Sufu plays an essential role in ciliary trafficking of Gli2 and Gli3. Sufu and Gli2 or Gli3 relocate to the tip of primary cilia in cells stimulated with a Hh ligand or a Smo agonist [105–107]. Kinesin family member 7 (Kif7) appears to serve a similar role but which is not redundant with Sufu [108]. Presumably inside the cilia Sufu and Kif7 dissociate and free Gli2 and Gli3, in their fully activated form leave the cilium and translocate directly to the nucleus, which is normally juxtaposed to the basal body [109]. In Sufu-deficient cells, processing of Gli2 and Gli3 is reduced, leading to ligand-independent activation of transcription [103]. However, maximal activation of Gli, by a still undetermined posttranslational modification, does not occur in the absence of Sufu. It is believed that Sufu has negative and positive functions in the Hh pathway and that the positive function reflects its role in promoting Gli2 and Gli3 accumulation in the cilium.

7 Gli Transcription Factors

7.1 Domain Structure of Gli1, Gli2, and Gli3

The three Gli isoforms belong to the Gli-Kruppel family of zinc-finger containing transcription factors. Hh induces Gli1, while Gli2 and Gli3 are constitutively expressed and function both as repressors or activators, depending on the level of pathway activation. They all contain a C-terminal transactivator domain, five C_2H_2 Kruppel-type Zn^{2+} finger motifs that form the DNA-binding domain, and a Sufu binding site in the N-terminal region [109, 110]. The DNA binding domain recognizes the sequence 5′- GACCACCCA [110]. In addition, they have a phosphorylation site cluster downstream of the Zn-fingers that contain between 4 (Gli1) and 16 (Gli2 and 3) phosphorylation sites, and Gli2 and Gli3 have an N-terminal repressor domain that is absent in Gli1.

Gli/Ci proteins possess a nuclear localization signal (NLS) and a nuclear export signal (NES), both of which are key signatures for controlling nucleocytoplasmic shuttling. The NLS of the Gli/Ci proteins has been mapped adjacent to the fifth Zn-finger domain. It contains two clusters of basic residues (classical bipartite-type), which are conserved in Gli/Ci homologues. The two basic clusters were predicted to fit in the two binding interfaces of Importin α [111]. Proteolytic processing of Gli2 and Gli3 eliminates the NES, promoting nuclear retention of GliRs. In addition, phosphorylation by PKA of a cluster near the NLS appears to oppose nuclear translocation of full-length Gli2 and Gli3.

7.2 Regulation of Gli Activity by Phosphorylation

PKA phosphorylates Gli2/3 at six conserved Ser residues (P1–6) immediately C-terminal of the Zn-finger domain [112, 113]. The phosphorylation of the first four of these residues (P1–4) by PKA initiates a cascade phosphorylation by GSK3 and CK1 that leads to the partial processing of full-length Glis into GliR fragments by the proteasome. Interestingly, loss of phosphorylation at sites P1–4 is not sufficient for full activation of Gli2 and Gli3 [114]. The phenotype of transgenic mice expressing non-phosphorylatable mutations in P1–4 of Gli2 does not seem to indicate full activation of Gli2, although it could be explained by retention of Gli3R formation [115]. However, it was recently shown that all three Gli isoforms contain an additional large cluster of phospho-sites, termed Pc-g, which do not fully conform to consensus PKA phosphorylation sites, and that are located N-terminal of the Zn-finger domain [114]. Activation of Smo leads to remodeling of the Gli phosphorylation pattern, with the P1-6 PKA target sites dephosphorylated, while the Pc-g cluster undergoes phosphorylation by unidentified kinase(s) [114]. It is highly probable that differential phosphorylation serves to decode increasing levels of Hh signaling, particularly during embryonic development when Hh proteins act as morphogens.

7.3 Partial Degradation of Gli2 and Gli3

The ubiquitin-proteasome system modulates Gli2 and Gli3 activity in two ways: it completely degrades the fully activated transcription factors in the presence of Hh and partially degrades Gli2 and Gli3 to convert them to repressors in the absence of Hh [116]. While the most common function of the proteasome is to degrade proteins completely, partial processing occurs when the proteasome runs into a folded domain that stalls its progress along the polypeptide, giving the remainder of the protein a chance to escape degradation. This type of proteasomal-mediated processing is rare and has been also described for the Nuclear Factor-κB (NF-κB) p50 subunit [117].

Gli3 is very efficiently processed into Gli3R, not so Gli2. A 200 amino acids region C-terminal of the Zn-finger domain was identified as the processing determinant domain (PDD) [118]. Transplanting the PDD of Gli3 into Gli2 was able to transfer the ability to be processed efficiently. However, it was later established that the PDD was not sufficient to induce partial proteasomal degradation of Gli1, and further studies revealed that the processing signal is encoded in three distinct regions of Gli3: the Zn-finger domain, a linker region, and a degron sequence [116].

Proteasomal recognition of Gli2 and Gli3 is mediated by ubiquitination. SCF E3 ubiquitin ligases are composed of Skp1, Cul1, Roc1, and F-box proteins, where the F-box proteins are the substrate recognizing subunits. The β-transducin repeat-containing F-box protein (β-TrCP) has been identified as the $SCF^{\beta\text{-TrCP}}$ E3 ubiquitin ligase responsible for Gli2 degradation. It directly binds to and directs ubiquitination of Gli2 [119]. In addition to β-TrCP, the adaptor protein Speckle-type PDZ protein (SPOP) also interacts with Gli2 and Gli3 and promotes proteasome-dependent degradation by the SPOP-Cul3 E3 ligase complex [104, 120].

As mentioned at the beginning of this section, Gli1 is not processed by the proteasome but it is completely degraded by it. It was recently reported that the adaptor protein Numb, which targets substrates for ubiquitination by recruiting the HECT-type E3 ligase Itch/AIP4, interacts with Gli1 [121]. Both Numb and Itch/AIP4 were shown to increase Gli1 polyubiquitination and full degradation, effectively functioning as negative regulators of Hh signaling [121]. Another pathway of Gli1 degradation in response to genotoxic stress is through the histone acetyltransferase (HAT) p300/CBP-associated factor (PCAF), which displays ubiquitin transferase activity in addition to its better-known acetyltransferase activity. PCAF induces the ubiquitination and degradation of Gli1, demonstrating that PCAF is a novel E3-ubiquitin ligase for Gli1 [122].

7.4 Transcriptional Activity of Gli Isoforms

7.4.1 Coactivators

Despite PCAF being a negative regulator of Gli1 stability, a recent study reported that PCAF is a positive regulator of the Hh pathway through its acetyltransferase activity [123]. PCAF acts as a coactivator that is recruited to the region surrounding the transcription start site of target genes. Both the level of Hh pathway activation and the amount of acetylated histone 3 in the Lys9 residue (H3K9) associated with the *Gli1* and *Ptch1* promoters were reduced by siRNA-mediated depletion of PCAF in NIH3T3 cells [123]. These findings suggest that acetylation of histone 3 (H3) in Lys9 is a requirement for Gli2 and Gli3 transcriptional activity.

Gli1 and Gli2 undergo acetylation/deacetylation as a central transcriptional checkpoint of Hh signaling, where deacetylation promotes transcriptional activity [124]. Class I HDAC depletion or their pharmacological inhibitors reduce Gli-luciferase activity in SAG-treated cells, and HDAC1 and HDAC2 overexpression potentiate Gli1 and Gli2 transcriptional activity, but not that of Gli3 [124]. ChIP analysis showed that HDAC1 and Gli1 form a complex recruited at the *Ptch1* promoter region. Activation of Gli1 is not linked to histone deacetylation, but rather to deacetylation of Gli1 and Gli2 by HDAC1. Gli1 is acetylated in Lys518, while Gli2 is acetylated in Lys757 [124, 125]. The HAT coactivator p300 acetylates Gli2 at the conserved Lys757 inhibiting Hh target gene expression by preventing Gli2 entry into chromatin [125]. Gli1 and Gli2 are deacetylated by a protein complex containing HDAC and an E3 ubiquitin ligase formed by Cullin3 and REN^{KCTD11} to stimulate Gli-target gene transcription [124].

Gli3 contains an N-terminal CREB-binding protein (CBP)-binding domain, and physical association with CBP is required for Gli3-mediated transcriptional activation [126]. Another coactivator of Gli3 is MED12, one of the subunits of Mediator, an evolutionarily conserved multiprotein interface between gene-specific transcription factors and RNA polymerase II. MED12 interacts with Gli3 and serves to reverse Mediator-dependent suppression of Shh target gene transcription [127].

7.4.2 Chromatin Remodeling by co-Repressors

Histone deacetylases (HDACs) remove acetyl groups from histone lysine residues allowing them to interact more strongly with DNA and preventing access of the transcriptional machinery. The inhibitory function of Gli3 is mediated by recruitment of HDACs to the vicinity of promoters of Gli-target genes. The co-repressor Ski is required for Gli3R and Gli2R-mediated repression, and can also inhibit transcriptional activation by the full-length form of Gli3 [128]. Ski binds to a conserved region in the N-terminal domain of Gli2 and Gli3 and is necessary for their association with HDACs. Sufu can also modulate the chromatin landscape of the Gli transcription factors. Sufu has been described to interact with Sin3A-Associated Protein (SAP18), which allows the formation of a Sufu/Gli repressor complex with mSin3-HDACs to promoters containing the Gli-binding elements [129].

7.4.3 Regulation of Gli Activity by Hedgehog-Independent Pathways

The past decade has witnessed an explosion of non-canonical ways of activation of Gli-dependent transcription, which is highly relevant for many cancer types that display high Gli1 expression despite absence of mutations in the upstream components of the Hh pathway. For instance, Transforming Growth Factor-β (TGF-β) signaling was shown to induce expression of Gli2 in fibroblasts, keratinocytes, and mammary epithelial cells in a Smad3-dependent manner [130]. This signaling crosstalk occurred downstream of Smo, as addition of a Smo antagonist was unable to prevent induction of Gli2 expression. Promoter analyses of the *Gli2* gene confirmed the presence of a Smad-responsive element [131].

Regulation of Gli activity by the Mitogen Activated Protein (MAP) kinase signaling pathway was first suggested by positive effects of FGF in Shh-dependent processes such as limb bud development and dopaminergic, serotoninergic, and oligodendrocyte specification [132–134]. Later, it was shown that the Mitogen-activated Protein Kinase (MEK)/Extracellular Regulated Kinase (Erk) MAP kinases promote activation of a Gli-luciferase reporter gene by acting on the N-terminal region of Gli1, which contains consensus MAP kinase phosphorylation sites [135, 136].

GPCR-mediated signals can also positively regulate Gli transcription. An activated mutant of the heterotrimeric G protein subunit $G\alpha_{13}$ was shown to stimulate Gli transcriptional activity in C310T1/2, MC3T3, and several pancreatic cancer cells, but not in fibroblasts [137]. Like for TGF-β, $G\alpha_{13}$ activation of Gli activity is independent of Smo. Rather, a constitutively active mutant of $G\alpha_{13}$ ($G\alpha_{13}$QL) acts partly via p38 MAP kinase and the Tec family of soluble tyrosine kinases to increase Gli activation. Moreover, this study demonstrated that an agonist for a GPCR coupled to G_{13} (CCK-A) is able to stimulate Gli activity through endogenous $G\alpha_{13}$ [137].

8 Non-Canonical Hedgehog Pathways

8.1 Non-Canonical Hedgehog Signaling Type I

Patched-1 has been classified as a dependence receptor, meaning that cell survival is dependent on the presence of the ligand when the receptor is expressed [38, 138]. In the absence of Hh ligands, overexpression of Patched-1 induces apoptosis both in 293 T cells in culture and in neuroepithelial cells in the chick neural tube [38]. Cleavage of Patched-1 by caspases at the C-terminal tail is essential for induction of apoptosis and a truncated Patched-1 mutant (Ptch1;1–1,392) induces cell death regardless of the presence of Shh [38]. This process is independent of the downstream elements of the canonical Hh pathway, as overexpression of Smo cannot prevent cell death. In line with those findings, we have found that depletion of Patched-1 in endothelial cells extends cell survival after serum withdrawal and reduces caspase-3 activation [139]. Importantly, while Shh increased endothelial cell survival, its effect was insensitive to Smo inhibitors and could not be mimicked by Smo agonists such as SAG, indicating that inhibition of the Patched-1 pro-apoptotic function is independence of Smo [139]. Later studies revealed that Patched-1 induces apoptosis by recruitment of a pro-apoptotic complex that includes caspase-9, Downregulated in Rhabdomyosarcoma LIM Protein (DRAL), and TUCAN-1 to the C-terminal tail (Fig. 3) [39].

The importance of the C-tail of Patched-1 was also evident by the finding that ubiquitination of a single Lys residue in the region of the C-tail that is clipped off by caspase is necessary for internalization and degradation of Patched-1, and that its mutation increases Patched-1 apoptotic potential (X.L. Chen et al., unpublished observations).

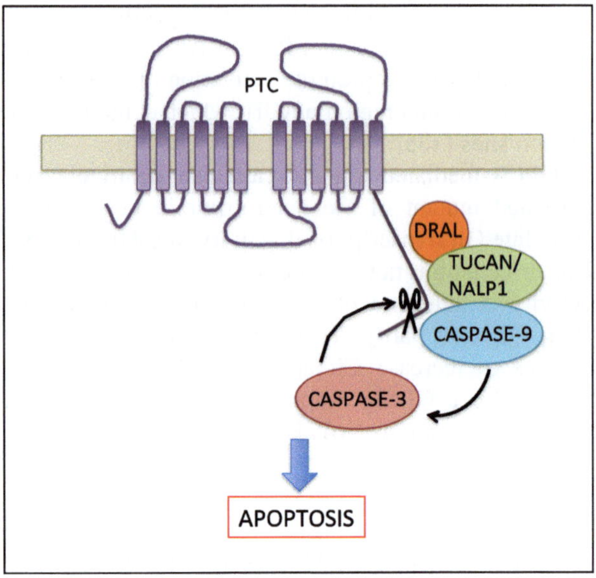

Fig. 3 Non-canonical Hedgehog signaling Type I. In the absence of a Hh ligand, Patched1 (PTC) recruits a pro-apoptotic complex to its C-terminal domain that contains the adaptor proteins DRAL and TUCAN/NALP1 and caspase-9. Activation of caspase-9 is followed by caspase-3 activation, which further amplifies the cell death signal by cleaving a region of Patched1 that regulates its apoptotic potential

8.2 Non-Canonical Hedgehog Signaling Type II

Activation of Smo serves additional roles independently of its central requirement for activation of the Gli transcription factors. Non-canonical Hh signaling type II mediates an increasing number of cellular responses that are sensitive to Smo agonists and inhibitors (Fig. 4). Importantly, all known pathways activated downstream of Smo appear to be mediated by the heterotrimeric G protein G_i.

8.2.1 Regulation of the Actin Cytoskeleton

Activation of Smo by Shh-elicited Patched-1 de-repression or directly with Smo small molecule agonists has profound effects on the actin cytoskeleton of fibroblasts, stellate cells, neurons, and endothelial cells [84, 139–147]. The cytoskeletal changes precede cell motility, morphological changes or axon guidance. In fibroblasts, Smo promotes motility and migration, both random and directional toward a gradient of Shh [84, 140, 141, 144]. Migration of fibroblasts in response to Shh occurs rapidly, incompatible with transcription, and is not inhibited by Gli3R [84, 140]. Activation of Smo in those cells engages the small GTPases RhoA and Rac1 by a mechanism requiring G_i proteins and PI-3 kinase, and does not require the primary cilium nor trafficking of Smo to the primary cilium [84, 140, 144]. Modulation of the actin cytoskeleton by Hh ligands has also been reported in endothelial cells in culture [139]. All three Hh ligands promote stress fiber formation in endothelial cells and stimulate tubulogenesis in 3D cultures [139, 145]. Importantly, RhoA and Rac1 are also activated by the three Hh proteins in a

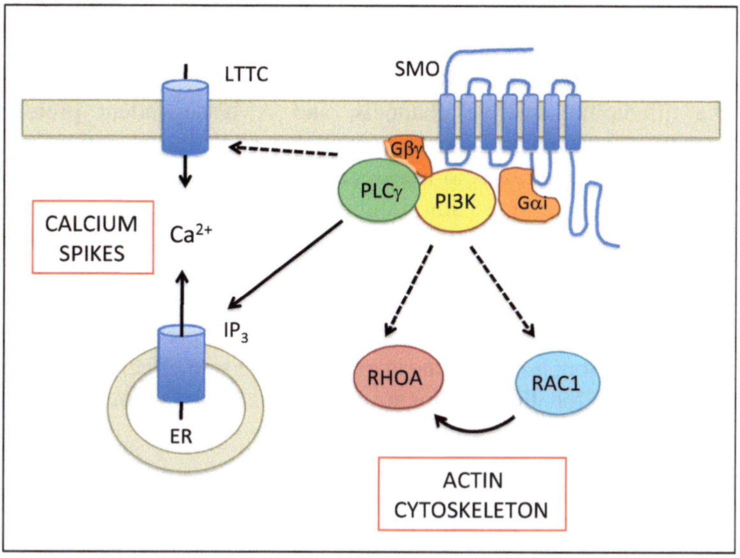

Fig. 4 Non-canonical Hedgehog signaling Type II. Activation of Smoothened leads to dissociation of G_i into Gαi and Gβγ, activation of PI3K and the small GTPases RhoA and Rac1, which then modulate the actin cytoskeleton and stimulate phospholipase C-γ (PLCγ) leading to a raise in intracellular calcium from both L-type calcium channel (LTCC) and the endoplasmic reticulum (ER)

similar manner than fibroblasts and are necessary mediators of tubulogenesis [139]. In stellate cells, Ihh enhances migration by a Gli1-independent pathway [146]. Moreover, in cholangiocarcinoma cell lines, which lack primary cilia and present a marked reduction of canonical Hh signaling, direct activation of Smo with purmorphamine induced remodeling of the actin cytoskeleton with formation of filopodia and lamellipodia [147]. As described for other cell types, migration and actin cytoskeleton remodeling were sensitive to inhibition of G_i proteins [147]. In retinal ganglion neurons in the chick embryo, addition of Shh induces axon growth cone retraction, and this effect is mediated by a decrease in cAMP within minutes of Shh addition, also supporting a role of G_i proteins in the regulation of this process [148].

8.2.2 Regulation of Calcium Levels

Activation of Smo by Shh or SAG in spinal neuron precursors leads to increased spontaneous calcium spike activity through a nifedipine-sensitive calcium channel, which pharmacologically is likely an L-type calcium channel (LTCC) [149]. The use of pertussis toxin was essential to conclude that G_i proteins mediate this process. Another report indicated that non-canonical Hh signaling participates in regulation of metabolism [150]. In the study, activation of Smo in adipocytes or myotubes by Shh, purmorphamine, SAG, or the oxysterols 25OHC and 20(*S*)OHC caused a rapid increase in intracellular calcium, followed by glucose uptake through Glucose transporter-4 (GLUT-4) via sequential activation of G_i proteins, LTCC-like nifedipine-sensitive channels, and AMP-dependent protein kinase (Ampk) [150]. Contrary to signaling to Rac1 and RhoA, activation of this calcium-Ampk axis requires Smo translocation to the primary cilium. Consistent with this notion, a subgroup of Smo inhibitors that relocalize Smo to the primary cilium (cyclopamine, GDC-0449) but not Smo inhibitors that prevent ciliary translocation (SANT-1, LDE-225, XL-139) promoted calcium influx and glucose uptake [150]. The novel function of some canonical Hh pathway inhibitors as agonists of this type of non-canonical signaling is unprecedented and requires further characterization; however, it can explain some side effects of Smo inhibitors used in the clinic for the treatment of cancer, such as muscle spams and wasting [151].

8.2.3 Other Roles of G_i Proteins in non-Canonical Signaling Type II

As mentioned in the previous sections, numerous evidences support a key role of Smo/G_i coupling in activation of RhoA and Rac1 and stimulation of calcium uptake by non-canonical Hh signaling type II. Another example of the physiological relevance of activation of G_i proteins by Smo is during activation of the NF-κB transcription factor in diffuse B-cell lymphoma cells [152]. In those cells, activation of Smo augments its physical interaction with G_i proteins and promotes the recruitment of a signaling complex containing Caspase Recruitment Domain-Containing Protein 11 (CARMA), B cell Lymphoma-10 (Bcl-10), and Mucosa-Associated Lymphoid Tissue-1 (MALT1), which in turn initiate activation of p65/p50 NF-κB. The mechanism is rapid and independent of Gli-dependent transcription. Remarkably, this study and a previous one also showed evidences of activation of G_{12} by Smo [152, 153].

Altogether, numerous evidences demonstrate that Smo does couple to heterotrimeric G proteins of the G_i family. The significance of that interaction is more obvious in non-canonical Hh signaling, but it seems to be permissive for Gli activation in some cell types, possibly due to the consequent drop in cAMP levels and reduction of PKA activity, a strong negative regulator of Gli stability.

8.2.4 Discovery of Smo Biased Agonists

A computer-aided drug design based on the structures of purmorphamine and SAG, two potent Smo agonists, identified the quinolinonecarboxamide compound GSA-10 as a predicted Smo ligand [154]. GSA-10 induced osteoblast differentiation of C3H10T1/2 cells with micromolar potency in a Smo-dependent manner. Remarkably, several antagonists of Smo (LDE-225, SANT-1, and M25) can prevent GSA-10 effects, but not cyclopamine or CUR61414, suggesting that the binding pocket of

GSA-10 and other Smo agonists differs and that at least two distinct active conformations of Smo exist. However, as opposed to purmorphamine and SAG, GSA-10 failed to activate Gli transcription, as determined by Gli-luciferase assays and by proliferation of cerebellar granule precursor cells, and did not induce Smo translocation into the primary cilium [154]. It is unclear if this compound targets Smo selectively, but it is an exciting starting point for the discovery of Smo biased agonists capable of activating selective branches of the Hh signaling pathway.

9 Final Remarks

The vertebrate Hh signaling pathway appears to have acquired additional functions beyond activation of the Ci/Gli transcription factors. The discovery of non-transcriptional functions of Hh signaling, including those related to survival by inhibition of a pro-apoptotic activity of Patched-1 and those related to G_i protein-mediated changes in actin cytoskeleton and calcium and glucose uptake, revealed a landscape of previously neglected biological responses to Hh that are critical for the analysis and understanding of pathway-based therapeutic development. The current preferred target for canonical Hh pathway inhibition is Smo. Many inhibitors that prevent Gli activation have been developed, one is FDA approved for the treatment of locally advanced and metastatic basal cell carcinoma and Shh-subtype medulloblastoma (GDC-0449, also known as Vismodegib from Genentech) and others are being evaluated in clinical trials (NVP-LDE-225/ Sonidegib, and NVP-LEQ-506 from Novartis, and BMS-833923/XL-139 from Bristol-Myers Squibb). However, the same compounds increase calcium uptake and glucose utilization by activation of a type of non-canonical signaling [150]. Thus, additional screening of the effect of Smo inhibitors on non-canonical signaling readouts, such as calcium uptake of lactate release, could be added during drug development to ensure that they do not act as biased agonists. Predictably, adapting the small molecule screenings to include those readouts will reduce the likelihood of the severe side effects that lead to patient-initiated drug withdrawal.

Further studies of vertebrate Hh signaling are desperately needed. Some big fundamental questions that remain unanswered include the nature of Patched-mediated Smo inhibition, the molecular nature of transmission of signals from activated Smo to Gli and Sufu, the posttranslational modification(s) of Gli2 and Gli3 necessary for formation of the labile active state, and the extent and mechanisms of non-canonical Hh-initiated pathways as well as of other signals that modulate Gli activation independently of Smo.

References

1. Robbins DJ, Fei DL, Riobo NA (2012) The Hedgehog signal transduction network. Sci Signal 5(246):re6
2. Brennan D, Chen X, Cheng L, Mahoney M, Riobo NA (2012) Noncanonical Hedgehog signaling. Vitam Horm 88:55–72
3. Nozawa YI, Lin C, Chuang PT (2013) Hedgehog signaling from the primary cilium to the nucleus: an emerging picture of ciliary localization, trafficking and transduction. Curr Opin Genet Dev 23(4):429–437
4. Briscoe J, Thérond PP (2013) The mechanisms of Hedgehog signalling and its roles in development and disease. Nat Rev Mol Cell Biol 14(7):416–429
5. Hall TM, Porter JA, Beachy PA, Leahy DJ (1995) A potential catalytic site revealed by the 1.7-A crystal structure of the amino-terminal signalling domain of Sonic hedgehog. Nature 378(6553):212–216
6. Fuse N, Maiti T, Wang B et al (1999) Sonic hedgehog protein signals not as a hydrolytic enzyme but as an apparent ligand for Patched. Proc Natl Acad Sci U S A 96(20):10992–10999
7. Hall TM, Porter JA, Young KE et al (1997) Crystal structure of a Hedgehog autoprocessing domain: homology between Hedgehog and self-splicing proteins. Cell 91(1):85–97
8. Lee JJ, Ekker SC, von Kessler DP et al (1994) Autoproteolysis in hedgehog protein biogenesis. Science 266(5190):1528–1537
9. Porter JA, von Kessler DP, Ekker SC et al (1995) The product of hedgehog autoproteolytic cleavage active in local and long-range signalling. Nature 374(6520):363–366
10. Porter JA, Ekker SC, Park WJ et al (1996) Hedgehog patterning activity: role of a lipophilic modification mediated by the carboxy-terminal autoprocessing domain. Cell 86(1):21–34
11. Porter JA, Young KE, Beachy PA (1996) Cholesterol modification of hedgehog signaling proteins in animal development. Science 274(5285):255–259
12. Tokhunts R, Singh S, Chu T et al (2010) The full-length unprocessed hedgehog protein is an active signaling molecule. J Biol Chem 285(4):2562–2568
13. Pepinsky RB, Zeng C, Wen D et al (1998) Identification of a palmitic acid-modified form of human Sonic hedgehog. J Biol Chem 273(22):14037–14045
14. Mann RK, Beachy PA (2004) Novel lipid modifications of secreted protein signals. Annu Rev Biochem 73:891–923
15. Farazi TA, Waksman G, Gordon JI (2001) The biology and enzymology of protein N-myristoylation. J Biol Chem 276:39501–39504
16. Buglino JA, Resh MD (2012) Palmitoylation of Hedgehog proteins. Vitam Horm 88:229–252
17. Buglino JA, Resh MD (2008) Hhat is a palmitoylacyltransferase with specificity for N-palmitoylation of Sonic Hedgehog. J Biol Chem 283:22076–22088
18. Chen MH, Li YJ, Kawakami T et al (2004) Palmitoylation is required for the production of a soluble multimeric Hedgehog protein complex and long-range signaling in vertebrates. Genes Dev 18:641–659
19. Martinez-Chinchilla P, Riobo NA (2008) Purification and bioassay of hedgehog ligands for the study of cell death and survival. Methods Enzymol 446:189–204
20. Caspary T, Garcia-Garcia MJ, Huangfu D et al (2002) Mouse Dispatched homolog1 is required for long-range, but not juxtacrine, Hh signaling. Curr Biol 12:1628–1632
21. Kawakami T, Kawcak T, Li YJ et al (2002) Mouse dispatched mutants fail to distribute hedgehog proteins and are defective in hedgehog signaling. Development 129:5753–5765
22. Ma Y, Erkner A, Gong R et al (2002) Hedgehog-mediated patterning of the mammalian embryo requires transporter-like function of dispatched. Cell 111:63–75
23. Creanga A, Glenn TD, Mann RK et al (2012) Scube/You activity mediates release of dually lipid-modified Hedgehog signal in soluble form. Genes Dev 26(12):1312–1325
24. Zeng X, Goetz JA, Suber LM et al (2001) A freely diffusible form of Sonic hedgehog mediates long-range signalling. Nature 411:716–720

25. Gallet A, Ruel L, Staccini-Lavenant L, Therond PP (2006) Cholesterol modification is necessary for controlled planar long-range activity of Hedgehog in Drosophila epithelia. Development 133:407–418
26. Martínez MC, Larbret F, Zobairi F et al (2006) Transfer of differentiation signal by membrane microvesicles harboring hedgehog morphogens. Blood 108(9):3012–3020
27. Soleti R, Benameur T, Porro C et al (2009) Microparticles harboring Sonic Hedgehog promote angiogenesis through the upregulation of adhesion proteins and proangiogenic factors. Carcinogenesis 30(4):580–588
28. Benameur T, Soleti R, Porro C et al (2010) Microparticles carrying Sonic hedgehog favor neovascularization through the activation of nitric oxide pathway in mice. PLoS One 5(9): e12688
29. Witek RP, Yang L, Liu R et al (2009) Liver cell-derived microparticles activate hedgehog signaling and alter gene expression in hepatic endothelial cells. Gastroenterology 136(1):320–330
30. Roy S, Hsiung F, Kornberg TB (2011) Specificity of Drosophila cytonemes for distinct signaling pathways. Science 332(6027):354–358
31. Bischoff M, Gradilla AC, Seijo I et al (2013) Cytonemes are required for the establishment of a normal Hedgehog morphogen gradient in Drosophila epithelia. Nat Cell Biol 15(11):1269–1281
32. Sanders TA, Llagostera E, Barna M (2013) Specialized filopodia direct long-range transport of SHH during vertebrate tissue patterning. Nature 497(7451):628–632
33. Tseng TT, Gratwick KS, Kollman J et al (1999) The RND permease superfamily: an ancient, ubiquitous and diverse family that includes human disease and development proteins. J Mol Microbiol Biotechnol 1(1):107–125
34. Strutt H, Thomas C, Nakano Y et al (2001) Mutations in the sterol-sensing domain of Patched suggest a role for vesicular trafficking in Smoothened regulation. Curr Biol 11:608–613
35. Johnson RL, Zhou L, Bailey EC (2002) Distinct consequences of sterol sensor mutations in Drosophila and mouse patched homologs. Dev Biol 242:224–235
36. Hime GR, Lada H, Fietz MJ et al (2004) Functional analysis in Drosophila indicates that the NBCCS/PTCH1 mutation G509V results in activation of smoothened through a dominant-negative mechanism. Dev Dyn 229(4):780–790
37. Lu X, Liu S, Kornberg TB (2006) The C-terminal tail of the Hedgehog receptor Patched regulates both localization and turnover. Genes Dev 20:2539–2551
38. Thibert C, Teillet MA, Lapointe F et al (2003) Inhibition of neuroepithelial patched-induced apoptosis by sonic hedgehog. Science 301(5634):843–846
39. Mille F, Thibert C, Fombonne J et al (2009) The Patched dependence receptor triggers apoptosis through a DRAL-caspase-9 complex. Nat Cell Biol 11(6):739–746
40. Lindstrom E, Shimokawa T, Toftgard R, Zaphiropoulos PG (2006) PTCH mutations: distribution and analyses. Hum Mutat 27:215–219
41. Taipale J, Cooper MK, Maiti T, Beachy PA (2002) Patched acts catalytically to suppress the activity of Smoothened. Nature 418:892–897
42. Denef N, Neubuser D, Perez L, Cohen SM (2000) Hedgehog induces opposite changes in turnover and subcellular localization of patched and smoothened. Cell 102:521–531
43. Yavari A, Nagaraj R, Owusu-Ansah E et al (2010) Role of lipid metabolism in smoothened derepression in hedgehog signaling. Dev Cell 19(1):54–65
44. Bijlsma MF, Spek CA, Zivkovic D et al (2006) Repression of smoothened by patched-dependent pro-vitamin D3 secretion. PLoS Biol 4(8):e232
45. Uhmann A, Niemann H, Lammering B et al (2011) Antitumoral effects of calcitriol in basal cell carcinomas involve inhibition of hedgehog signaling and induction of vitamin D receptor signaling and differentiation. Mol Cancer Ther 10(11):2179–2188
46. Tang JY, Xiao TZ, Oda Y et al (2011) Vitamin D3 inhibits hedgehog signaling and proliferation in murine Basal cell carcinomas. Cancer Prev Res (Phila) 4(5):744–751

47. Dormoy V, Béraud C, Lindner V et al (2012) Vitamin D3 triggers antitumor activity through targeting hedgehog signaling in human renal cell carcinoma. Carcinogenesis 33(11):2084–2093
48. DeBerardinis AM, Banerjee U, Miller M et al (2012) Probing the structural requirements for vitamin D3 inhibition of the hedgehog signaling pathway. Bioorg Med Chem Lett 22(14):4859–4863
49. Bidet M, Joubert O, Lacombe B et al (2011) The hedgehog receptor patched is involved in cholesterol transport. PLoS One 6(9):e23834
50. Corbit KC, Aanstad P, Singla V et al (2005) Vertebrate Smoothened functions at the primary cilium. Nature 437(7061):1018–1021
51. Rohatgi R, Milenkovic L, Scott MP (2007) Patched1 regulates hedgehog signaling at the primary cilium. Science 317(5836):372–376
52. Bailey EC, Milenkovic L, Scott MP et al (2002) Several PATCHED1 missense mutations display activity in patched1-deficient fibroblasts. J Biol Chem 277(37):33632–33640
53. Huang S, Zhang Z, Zhang C et al (2013) Activation of Smurf E3 ligase promoted by smoothened regulates hedgehog signaling through targeting patched turnover. PLoS Biol 11(11):e1001721
54. Tenzen T, Allen BL, Cole F et al (2006) The cell surface membrane proteins Cdo and Boc are components and targets of the Hedgehog signaling pathway and feedback network in mice. Dev Cell 10(5):647–656
55. Allen BL, Song JY, Izzi L et al (2011) Overlapping roles and collective requirement for the coreceptors GAS1, CDO, and BOC in SHH pathway function. Dev Cell 20(6):775–787
56. Kang JS, Gao M, Feinleib JL et al (1997) CDO: an oncogene-, serum-, and anchorage-regulated member of the Ig/fibronectin type III repeat family. J Cell Biol 138(1):203–213
57. McLellan JS, Zheng X, Hauk G et al (2008) The mode of Hedgehog binding to Ihog homologues is not conserved across different phyla. Nature 455(7215):979–983
58. McLellan JS, Yao S, Zheng X et al (2006) Structure of a heparin-dependent complex of Hedgehog and Ihog. Proc Natl Acad Sci U S A 103(46):17208–17213
59. Mullor JL, Ruiz i Altaba A (2002) Growth, hedgehog and the price of GAS. Bioessays 24(1):22–26
60. Allen BL, Tenzen T, McMahon AP (2007) The Hedgehog-binding proteins Gas1 and Cdo cooperate to positively regulate Shh signaling during mouse development. Genes Dev 21:1244–1257
61. Pineda-Alvarez DE, Roessler E, Hu P et al (2012) Missense substitutions in the GAS1 protein present in holoprosencephaly patients reduce the affinity for its ligand, SHH. Hum Genet 131(2):301–310
62. Bae GU, Domené S, Roessler E et al (2011) Mutations in CDON, encoding a hedgehog receptor, result in holoprosencephaly and defective interactions with other hedgehog receptors. Am J Hum Genet 89(2):231–240
63. Zhang W, Hong M, Bae GU et al (2011) Boc modifies the holoprosencephaly spectrum of Cdo mutant mice. Dis Model Mech 4(3):368–380
64. Kristiansen K (2004) Molecular mechanisms of ligand binding, signaling, and regulation within the superfamily of G-protein-coupled receptors: molecular modeling and mutagenesis approaches to receptor structure and function. Pharmacol Ther 103(1):21–80
65. Zhao Y, Tong C, Jiang J (2007) Hedgehog regulates smoothened activity by inducing a conformational switch. Nature 450(7167):252–258
66. Wang C, Wu H, Katritch V et al (2013) Structure of the human smoothened receptor bound to an antitumour agent. Nature 497(7449):338–343
67. Nachtergaele S, Whalen DM, Mydock LK et al (2013) Structure and function of the Smoothened extracellular domain in vertebrate Hedgehog signaling. Elife 2:e01340
68. Myers BR, Sever N, Chong YC et al (2013) Hedgehog pathway modulation by multiple lipid binding sites on the smoothened effector of signal response. Dev Cell 26(4):346–357

69. Nedelcu D, Liu J, Xu Y et al (2013) Oxysterol binding to the extracellular domain of Smoothened in Hedgehog signaling. Nat Chem Biol 9(9):557–564
70. Corcoran RB, Scott MP (2006) Oxysterols stimulate Sonic hedgehog signal transduction and proliferation of medulloblastoma cells. Proc Natl Acad Sci U S A 103(22):8408–8413
71. Dwyer JR, Sever N, Carlson M et al (2007) Oxysterols are novel activators of the hedgehog signaling pathway in pluripotent mesenchymal cells. J Biol Chem 282(12):8959–8968
72. Nachtergaele S, Mydock LK, Krishnan K et al (2012) Oxysterols are allosteric activators of the oncoprotein Smoothened. Nat Chem Biol 8(2):211–220
73. Chen Y, Li S, Tong C et al (2010) G protein-coupled receptor kinase 2 promotes high-level Hedgehog signaling by regulating the active state of Smo through kinase-dependent and kinase-independent mechanisms in Drosophila. Genes Dev 24(18):2054–2067
74. Chen W, Ren XR, Nelson CD et al (2004) Activity-dependent internalization of smoothened mediated by beta-arrestin 2 and GRK2. Science 306(5705):2257–2260
75. Wilbanks AM, Fralish GB, Kirby ML et al (2004) Beta-arrestin 2 regulates zebrafish development through the hedgehog signaling pathway. Science 306(5705):2264–2267
76. Kovacs JJ, Whalen EJ, Liu R et al (2008) Beta-arrestin-mediated localization of smoothened to the primary cilium. Science 320(5884):1777–1781
77. Shi Q, Li S, Jia J, Jiang J (2011) The Hedgehog-induced Smoothened conformational switch assembles a signaling complex that activates Fused by promoting its dimerization and phosphorylation. Development 138(19):4219–4231
78. Lum L, Zhang C, Oh S et al (2003) Hedgehog signal transduction via Smoothened association with a cytoplasmic complex scaffolded by the atypical kinesin, Costal-2. Mol Cell 12(5):1261–1274
79. Chen Y, Sasai N, Ma G et al (2011) Sonic Hedgehog dependent phosphorylation by CK1α and GRK2 is required for ciliary accumulation and activation of smoothened. PLoS Biol 9(6):e1001083
80. Xia R, Jia H, Fan J et al (2012) USP8 promotes smoothened signaling by preventing its ubiquitination and changing its subcellular localization. PLoS Biol 10(1):e1001238
81. Li S, Chen Y, Shi Q et al (2012) Hedgehog-regulated ubiquitination controls smoothened trafficking and cell surface expression in Drosophila. PLoS Biol 10(1):e1001239
82. Riobo NA, Saucy B, Dilizio C, Manning DR (2006) Activation of heterotrimeric G proteins by Smoothened. Proc Natl Acad Sci U S A 103(33):12607–12612
83. Shen F, Cheng L, Douglas AE et al (2013) Smoothened is a fully competent activator of the heterotrimeric G protein G(i). Mol Pharmacol 83(3):691–697
84. Polizio AH, Chinchilla P, Chen X et al (2011) Heterotrimeric Gi proteins link Hedgehog signaling to activation of Rho small GTPases to promote fibroblast migration. J Biol Chem 286(22):19589–19596
85. Ogden SK, Fei DL, Schilling NS et al (2008) G protein Galphai functions immediately downstream of Smoothened in Hedgehog signalling. Nature 456(7224):967–970
86. Hammerschmidt M, McMahon AP (1998) The effect of pertussis toxin on zebrafish development: a possible role for inhibitory G-proteins in hedgehog signaling. Dev Biol 194(2):166–171
87. Low WC, Wang C, Pan Y et al (2008) The decoupling of Smoothened from Galphai proteins has little effect on Gli3 protein processing and Hedgehog-regulated chick neural tube patterning. Dev Biol 321(1):188–196
88. Barzi M, Kostrz D, Menendez A, Pons S (2011) Sonic Hedgehog-induced proliferation requires specific Gα inhibitory proteins. J Biol Chem 286(10):8067–8074
89. Preat T (1992) Characterization of Suppressor of fused, a complete suppressor of the fused segment polarity gene of Drosophila melanogaster. Genetics 132:725–736
90. Cooper AF, Yu KP, Brueckner M et al (2005) Cardiac and CNS defects in a mouse with targeted disruption of suppressor of fused. Development 132:4407–4417

91. Sv¨ard J, Heby-Henricson K, Persson-Lek M et al (2006) Genetic elimination of Suppressor of fused reveals an essential repressor function in the mammalian Hedgehog signaling pathway. Dev Cell 10:187–197
92. Kogerman P, Grimm T, Kogerman L et al (1999) Mammalian suppressor-of-fused modulates nuclear-cytoplasmic shuttling of Gli-1. Nat Cell Biol 1(5):312–319
93. Ding Q, Si F, Meng X et al (1999) Mouse suppressor of fused is a negative regulator of sonic hedgehog signaling and alters the subcellular distribution of Gli1. Curr Biol 9(19):1119–1122
94. Zhang Y, Fu L, Qi X et al (2013) Structural insight into the mutual recognition and regulation between Suppressor of Fused and Gli/Ci. Nat Commun 4:2608
95. Cherry AL, Finta C, Karlström M et al (2013) Structural basis of SUFU-GLI interaction in human Hedgehog signalling regulation. Acta Crystallogr D Biol Crystallogr 69(12):2563–2579
96. M´ethot N, Basler K (2000) Suppressor of Fused opposes Hedgehog signal transduction by impeding nuclear accumulation of the activator form of Cubitus interruptus. Development 127:4001–4110
97. Wang QT, Holmgren RA (2000) Nuclear import of Cubitus interruptus is regulated by Hedgehog via mechanism distinct from Ci stabilization and Ci activation. Development 127:3131–3139
98. Wang G, Amanai K, Jiang J et al (2000) Interactions with Costal2 and Suppressor of fused regulate nuclear translocation and activity of Cubitus interruptus. Genes Dev 14:2893–2905
99. Barnfield PC, Zhang X, Thanabalasingham V et al (2005) Negative regulation of Gli1 and Gli2 activator function by Suppressor of fused through multiple mechanisms. Differentiation 73:397–405
100. Murone M, Luoh SM, Stone D et al (2000) Gli regulation by the opposing activities of Fused and Suppressor of Fused. Nat Cell Biol 2:310–312
101. Merchant M, Vajdos FF, Ultsch M et al (2004) Suppressor of fused regulates Gli activity through a dual binding mechanism. Mol Cell Biol 24(19):8627–8641
102. Evangelista M, Lim TY, Lee J et al (2008) Kinome siRNA screen identifies regulators of ciliogenesis and Hedgehog signal transduction. Sci Signal 1:ra7
103. Chen MH, Wilson CW, Li YJ et al (2009) Cilium-independent regulation of Gli protein function by Sufu in Hedgehog signaling is evolutionarily conserved. Genes Dev 23(16):1910–1928
104. Wang C, Pan Y, Wang B (2010) Suppressor of fused and Spop regulate the stability, processing and function of Gli2 and Gli3 full-length activators but not their repressors. Development 137:2001–2009
105. Wen X, Lai CK, Evangelista M et al (2010) Kinetics of Hedgehog-dependent full-length Gli3 accumulation in primary cilia and subsequent degradation. Mol Cell Biol 30:1910–1922
106. Humke EW, Dorn KV, Milenkovic L et al (2010) The output of Hedgehog signaling is controlled by the dynamic association between Suppressor of Fused and the Gli proteins. Genes Dev 24(7):670–682
107. Tukachinsky H, Lopez LV, Salic A (2010) A mechanism for vertebrate Hedgehog signaling: recruitment to cilia and dissociation of SuFu-Gli protein complexes. J Cell Biol 191(2):415–428
108. Cheung HO, Zhang X, Ribeiro A et al (2009) The kinesin protein Kif7 is a critical regulator of Gli transcription factors in mammalian hedgehog signaling. Sci Signal 2(76):ra29
109. Hui CC, Angers S (2011) Gli proteins in development and disease. Annu Rev Cell Dev Biol 27:513–537
110. Kinzler KW, Vogelstein B (1990) The GLI gene encodes a nuclear protein which binds specific sequences in the human genome. Mol Cell Biol 10(2):634–642
111. Hatayama M, Aruga J (2012) Gli protein nuclear localization signal. Vitam Horm 88:73–89
112. Wang B, Fallon JF, Beachy PA (2000) Hedgehog-regulated processing of Gli3 produces an anterior/posterior repressor gradient in the developing vertebrate limb. Cell 100(4):423–434
113. Riobó NA, Lu K, Ai X et al (2006) Phosphoinositide 3-kinase and Akt are essential for Sonic Hedgehog signaling. Proc Natl Acad Sci U S A 103(12):4505–4510

114. Niewiadomski P, Kong JH, Ahrends R et al (2014) Gli protein activity is controlled by multisite phosphorylation in vertebrate Hedgehog signaling. Cell Rep 6(1):168–181
115. Pan Y, Wang C, Wang B (2009) Phosphorylation of Gli2 by protein kinase A is required for Gli2 processing and degradation and the Sonic Hedgehog-regulated mouse development. Dev Biol 326(1):177–189
116. Schrader EK, Harstad KG, Holmgren RA, Matouschek A (2011) A three-part signal governs differential processing of Gli1 and Gli3 proteins by the proteasome. J Biol Chem 286(45):39051–39058
117. Palombella VJ, Rando OJ, Goldberg AL, Maniatis T (1994) The ubiquitin-proteasome pathway is required for processing the NF-kappa B1 precursor protein and the activation of NF-kappa B. Cell 78(5):773–785
118. Pan Y, Wang B (2007) A novel protein-processing domain in Gli2 and Gli3 differentially blocks complete protein degradation by the proteasome. J Biol Chem 282(15):10846–10852
119. Bhatia N, Thiyagarajan S, Elcheva I et al (2006) Gli2 is targeted for ubiquitination and degradation by beta-TrCP ubiquitin ligase. J Biol Chem 281(28):19320–19326
120. Schwend T, Jin Z, Jiang K et al (2013) Stabilization of speckle-type POZ protein (Spop) by Daz interacting protein 1 (Dzip1) is essential for Gli turnover and the proper output of Hedgehog signaling. J Biol Chem 288(45):32809–32820
121. Di Marcotullio L, Greco A, Mazzà D et al (2011) Numb activates the E3 ligase Itch to control Gli1 function through a novel degradation signal. Oncogene 30(1):65–76
122. Mazzà D, Infante P, Colicchia V et al (2013) PCAF ubiquitin ligase activity inhibits Hedgehog/Gli1 signaling in p53-dependent response to genotoxic stress. Cell Death Differ 20(12):1688–1697
123. Malatesta M, Steinhauer C, Mohammad F et al (2013) Histone acetyltransferase PCAF is required for Hedgehog-Gli-dependent transcription and cancer cell proliferation. Cancer Res 73(20):6323–6333
124. Canettieri G, Di Marcotullio L, Greco A et al (2010) Histone deacetylase and Cullin3-REN (KCTD11) ubiquitin ligase interplay regulates Hedgehog signalling through Gli acetylation. Nat Cell Biol 12(2):132–142
125. Coni S, Antonucci L, D'Amico D et al (2013) Gli2 acetylation at lysine 757 regulates hedgehog-dependent transcriptional output by preventing its promoter occupancy. PLoS One 8(6):e65718
126. Dai P, Akimaru H, Tanaka Y et al (1999) Sonic Hedgehog-induced activation of the Gli1 promoter is mediated by GLI3. J Biol Chem 274(12):8143–8152
127. Zhou H, Kim S, Ishii S, Boyer TG (2006) Mediator modulates Gli3-dependent Sonic hedgehog signaling. Mol Cell Biol 26(23):8667–8682
128. Dai P, Shinagawa T, Nomura T et al (2002) Ski is involved in transcriptional regulation by the repressor and full-length forms of Gli3. Genes Dev 16(22):2843–2848
129. Cheng SY, Bishop JM (2002) Suppressor of Fused represses Gli-mediated transcription by recruiting the SAP18-mSin3 corepressor complex. Proc Natl Acad Sci U S A 99(8):5442–5447
130. Dennler S, André J, Alexaki I et al (2007) Induction of sonic hedgehog mediators by transforming growth factor-beta: Smad3-dependent activation of Gli2 and Gli1 expression in vitro and in vivo. Cancer Res 67(14):6981–6986
131. Dennler S, André J, Verrecchia F, Mauviel A (2009) Cloning of the human GLI2 Promoter: transcriptional activation by transforming growth factor-beta via SMAD3/beta-catenin cooperation. J Biol Chem 284(46):31523–31531
132. Laufer E, Nelson CE, Johnson RL et al (1994) Sonic hedgehog and Fgf-4 act through a signaling cascade and feedback loop to integrate growth and patterning of the developing limb bud. Cell 79(6):993–1003
133. Kessaris N, Jamen F, Rubin LL, Richardson WD (2004) Cooperation between sonic hedgehog and fibroblast growth factor/MAPK signalling pathways in neocortical precursors. Development 131(6):1289–1298

134. Ye W, Shimamura K, Rubenstein JL et al (1998) FGF and Shh signals control dopaminergic and serotonergic cell fate in the anterior neural plate. Cell 93(5):755–766
135. Riobo NA, Haines GM, Emerson CP Jr (2006) Protein kinase C-delta and mitogen-activated protein/extracellular signal-regulated kinase-1 control GLI activation in hedgehog signaling. Cancer Res 66(2):839–845
136. Whisenant TC, Ho DT, Benz RW et al (2010) Computational prediction and experimental verification of new MAP kinase docking sites and substrates including Gli transcription factors. PLoS Comput Biol 26:6(8)
137. Douglas AE, Heim JA, Shen F et al (2011) The alpha subunit of the G protein G13 regulates activity of one or more Gli transcription factors independently of smoothened. J Biol Chem 286(35):30714–30722
138. Bredesen DE, Mehlen P, Rabizadeh S (2004) Apoptosis and dependence receptors: a molecular basis for cellular addiction. Physiol Rev 84(2):411–430
139. Chinchilla P, Xiao L, Kazanietz MG, Riobo NA (2010) Hedgehog proteins activate pro-angiogenic responses in endothelial cells through non-canonical signaling pathways. Cell Cycle 9(3):570–579
140. Polizio AH, Chinchilla P, Chen X et al (2011) Sonic Hedgehog activates the GTPases Rac1 and RhoA in a Gli-independent manner through coupling of smoothened to Gi proteins. Sci Signal 4(200):pt7
141. Bijlsma MF, Borensztajn KS, Roelink H et al (2007) Sonic hedgehog induces transcription-independent cytoskeletal rearrangement and migration regulated by arachidonate metabolites. Cell Signal 19(12):2596–2604
142. Yam PT, Langlois SD, Morin S, Charron F (2009) Sonic hedgehog guides axons through a noncanonical, Src-family-kinase-dependent signaling pathway. Neuron 62(3):349–362
143. Yam PT, Kent CB, Morin S et al (2012) 14-3-3 proteins regulate a cell-intrinsic switch from sonic hedgehog-mediated commissural axon attraction to repulsion after midline crossing. Neuron 76(4):735–749
144. Bijlsma MF, Damhofer H, Roelink H (2012) Hedgehog-stimulated chemotaxis is mediated by smoothened located outside the primary cilium. Sci Signal 5(238):ra60
145. Kanda S, Mochizuki Y, Suematsu T et al (2003) Sonic hedgehog induces capillary morphogenesis by endothelial cells through phosphoinositide 3-kinase. J Biol Chem 278(10):8244–8249
146. Shinozaki S, Ohnishi H, Hama K et al (2008) Indian hedgehog promotes the migration of rat activated pancreatic stellate cells by increasing membrane type-1 matrix metalloproteinase on the plasma membrane. J Cell Physiol 216(1):38–46
147. Razumilava N, Gradilone SA, Smoot RL et al (2014) Non-canonical Hedgehog signaling contributes to chemotaxis in cholangiocarcinoma. J Hepatol 60(3):599–605
148. Trousse F, Martí E, Gruss P et al (2001) Control of retinal ganglion cell axon growth: a new role for Sonic hedgehog. Development 128(20):3927–3936
149. Belgacem YH, Borodinsky LN (2011) Sonic hedgehog signaling is decoded by calcium spike activity in the developing spinal cord. Proc Natl Acad Sci U S A 108(11):4482–4487
150. Teperino R, Amann S, Bayer M et al (2012) Hedgehog partial agonism drives Warburg-like metabolism in muscle and brown fat. Cell 151(2):414–426
151. Proctor AE, Thompson LA, O'Bryant CL (2014) Vismodegib: an inhibitor of the Hedgehog signaling pathway in the treatment of basal cell carcinoma. Ann Pharmacother 48(1):99–106
152. Qu C, Liu Y, Kunkalla K, Singh RR et al (2013) Trimeric G protein-CARMA1 axis links smoothened, the hedgehog receptor transducer, to NF-κB activation in diffuse large B-cell lymphoma. Blood 121(23):4718–4728
153. Kasai K, Takahashi M, Osumi N et al (2004) The G12 family of heterotrimeric G proteins and Rho GTPase mediate Sonic hedgehog signalling. Genes Cells 9(1):49–58
154. Gorojankina T, Hoch L, Faure H et al (2013) Discovery, molecular and pharmacological characterization of GSA-10, a novel small-molecule positive modulator of Smoothened. Mol Pharmacol 83(5):1020–1029

Top Med Chem (2015) 16: 43–108
DOI: 10.1007/7355_2014_61

Published online: 25 October 2014

Structure–Activity Relationships and Mechanism of Action of Small Molecule Smoothened Modulators Discovered by High-Throughput Screening and Rational Design

Fabrizio Manetti, Maurizio Taddei, and Elena Petricci

Abstract Smoothened (Smo) is the signal transducer of the Hedgehog (Hh) pathway and its stimulation or inhibition is considered a potential powerful tool in regenerative medicine and for the treatment of cancer. In the last years, many natural and nonnatural small molecules have been identified that are able to modulate the Hh pathway. Most of them target Smo, while only a few compounds are able to interact directly with upstream and downstream Hh pathway components. Although several compounds showed a remarkable potency and selectivity, their use induced emergence of mutated and resistant cell lines. In an attempt to find new chemical entries able to affect the Hh pathway and overcome limitation imposed by mutations and resistance, academic researchers and pharmaceutical companies are making further efforts to identify new drug-like small molecules to be included in the currently available therapeutic protocols for several types of cancers or in regenerative medicine and tissue repair.

Keywords Hedgehog signaling, Hit-to-lead compound optimization, Small molecule, Smoothened antagonists and agonists, Structure–activity relationships, Synthesis

Contents

F. Manetti (✉), M. Taddei, and E. Petricci
Department of Biotechnology, Chemistry and Pharmacy, University of Siena, via A. Moro 2, 53100 Siena, Italy
e-mail: fabrizio.manetti@unisi.it

Abbreviations

Adh7	Alcohol dehydrogenase 7
ADMETox	Absorption, distribution, metabolism, excretion, toxicity
AP	Alkaline phosphatase
APL	Acute promyelocytic leukemia
BCC	Basal cell carcinoma
BODIPY	Boron dimethylpyrrolydene
CML	Chronic myelogenous leukemia
CRD	Cysteine-rich domain
Cyp24A1	Cytochrome P24A1
Cyp450	Cytochrome P450
Dhh	Desert Hedgehog
DOS	Diversity-oriented synthesis
EC_{50}	Half-maximal effective concentration
FDA	Food and Drug Administration
FRET	Fluorescence resonance energy transfer
GANT	Glioma-associated oncogene homologue (Gli) antagonist
GCP	Granular cell precursor
GFP	Green fluorescent protein
GI_{50}	50% inhibition of cell growth
Gli1-3	Glioma-associated oncogene homologue 1–3
Gli-Luc assay	Glioma-associated oncogene homologue-luciferase assay
GPCR	G protein-coupled receptor
h	Hour
HEPM	Human embryonic palatal mesenchymal cells
hERG	Human ether-à-go-go-related gene
Hh	Hedgehog
Hh-Ag	Hedgehog agonist
Hh-Antag691	Hedgehog antagonist 691
HPI	Hedgehog pathway inhibitors
HTS	High-throughput screening
IC_{50}	Half-maximal inhibitory concentration
Ihh	Indian Hedgehog
KAAD-cyclopamine	3-keto-*N*-(aminoethyl-aminocaproyl-dihydrocinnamoyl) cyclopamine
LiHMDS	Lithium bis(trimethylsilyl)azanide
MAPK	Mitogen-activated protein kinase
MeI	Methyl iodide
mL	Milliliter
nM	Nanomolar
PBO	Piperonyl butoxide
PKC	Protein kinase C
PSA	Prostate specific antigen
Ptch	Patched
PXR	Pregnane X receptor

Ren	Retinoic acid, epidermal growth factor, nerve growth factor induced gene protein
SAG	Smoothened agonist
SANT	Smoothened antagonist
SAR	Structure–activity relationships
SFC	Supercritical fluid chromatography
Shh	Sonic Hedgehog
ShhN	N-terminal tail of sonic Hedgehog
SMANT	Smoothened mutant antagonist
Smo	Smoothened
SmoM2	Oncogenic Smoothened (Smo) mutant
SuFu	Suppressor of fused
TM	Transmembrane helix
VD2	Vitamin D2
VD3	Vitamin D3
VDR	Vitamin D receptor
μg	Microgram
μM	Micromolar

1 Introduction

Smoothened (Smo) is the signal transducer (the effector component) of the Hedgehog (Hh) signaling pathway that is also comprised of upstream ligands (Shh, Dhh, Ihh) and transmembrane receptors (Ptch1-2), as well as downstream regulators (i.e., SuFu, Ren) and transcription factors (Gli1-3). All of them orchestrate various aspects of embryonic pattern formation and adult tissue maintenance, in terms of remodeling, repair, and regeneration. However, deregulation of the Hh signaling pathway has been associated with a certain number of human cancers [1–3]. Therefore, targeting and modulating this pathway in cells has become a challenging issue and resulted in the identification of many classes of structurally different small molecules. Most of them specifically target Smo at different binding sites and are able to inhibit or enhance its activity. On the other hand, only a few small molecules showed different mechanisms of action, being able to target upstream or downstream components of the Hh pathway.

Recent literature is comprised of reviews describing small molecule inhibitors of the Hh signaling pathway [4–7], Smoothened inhibitors [8], Hh pathway agonists [9], and more general papers that attempt an exhaustive description of both agonists and antagonists targeting all the components of the pathway [10]. Following these efforts, we report here an up-to-date survey of small molecules that affect the Hh pathway: inhibitors targeting Smo at various binding sites; inhibitors without evidence of Smo binding; inhibitors targeting the upstream ShhN ligand or downstream transcription factors Gli1-2; agonists acting at Smo.

A particular emphasis is devoted to aspects of rational design and structure–activity relationships that have allowed identification of lead compounds and new anticancer drugs.

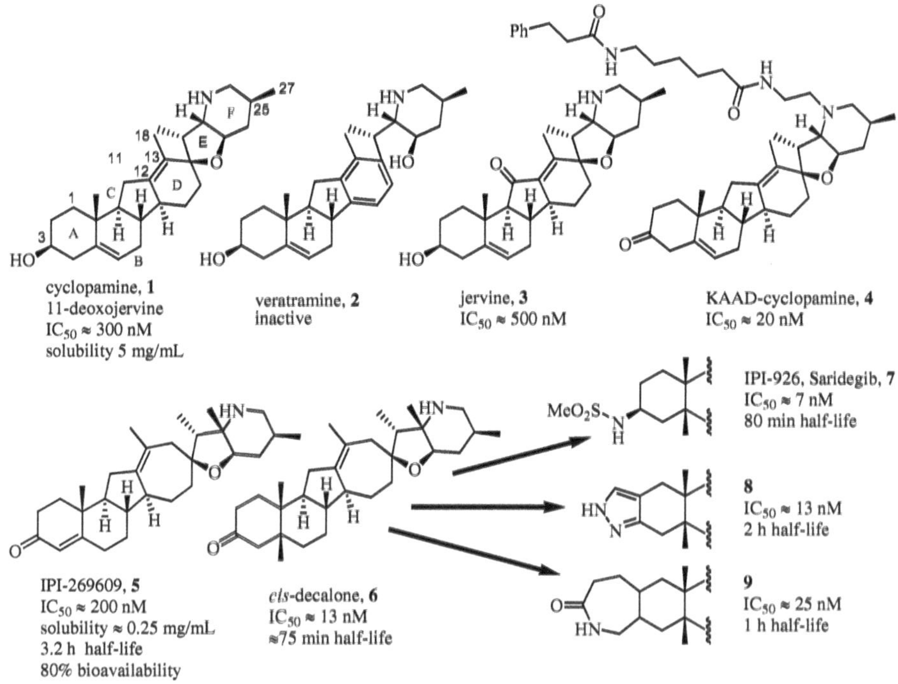

Fig. 1 Cyclopamine and its natural and nonnatural derivatives (the last discovered by Infinity Pharmaceuticals)

2 Compounds Affecting the Hh Signaling Pathway with Direct Evidence of Smo Binding at the Boron Dimethylpyrrolydene (BODIPY)-Cyclopamine Binding Site

2.1 Cyclopamine, Natural Derivatives, and Nonnatural Analogues (Infinity Pharmaceuticals)

Cyclopamine (**1**, Fig. 1) is a natural steroidal alkaloid isolated from the corn lily (*Veratrum californicum*, Melanthiaceae). In the past, it was deeply studied for its ability to induce holoprosencephaly and cyclopia in sheep. Cyclopamine (also referred to as 11-deoxojervine) directly binds to Smo, induces its accumulation in the primary cilium, and inhibits its activity by shifting Smo conformation to the closed form (for a description of the closed and open form of Smo, see Sect. 3.1) even if Hh ligands are present ($IC_{50} \approx 300$ nM). Given its activity in blocking the Hh pathway, cyclopamine was assayed as an antitumor agent in several xenograft models of cancer. Two major limitations prevented cyclopamine from being a possible anticancer drug: a poor aqueous solubility (about 5 μg/mL) and acid lability. In fact, by acid-catalyzed E-ring opening and subsequent D-ring aromatization, cyclopamine is rapidly converted into veratramine (**2**), inactive at the level of Hh pathway. Many natural and nonnatural cyclopamine derivatives have been described in the literature, showing comparable (jervine, **3**, $IC_{50} \approx 500$ nM) or better Hh inhibitory activity with respect to the parent compound. As an example, 3-keto-*N*-(aminoethyl-aminocaproyl-dihydrocinnamoyl)cyclopamine (KAAD-cyclopamine, **4**) [11] showed an activity more than ten-fold better ($IC_{50} \approx 20$ nM) than that of cyclopamine itself.

Selective delivery of cyclopamine to tumor tissues was attempted to reduce toxicity toward normal tissues and in part to overcome solubility and stability limitations. An example was represented by cyclopamine glucuronide used as selective inhibitor of U87MG glioblastoma cells. Twenty-eight hours after a combined treatment with β-glucuronidase, such a compound showed a 21 μM IC_{50} [12]. However, considering that the kinetics of drug delivery was limited by solubility of a de-glucuronidated intermediate bearing a phenolic moiety, a different cyclopamine glucuronide prodrug was synthesized by click chemistry with the insertion of a self-immolative linker between the sugar moiety and cyclopamine core. The corresponding de-glucuronidated phenolic intermediate was much more soluble and cyclopamine was fully released in the medium in less than two hours. While the new prodrug did not affect viability of U87 cells, combination with β-glucuronidase resulted in a 24 μM IC_{50}, comparable with that of cyclopamine alone (16 μM) [13]. In a similar attempt, cyclopamine was linked to specific amino acid sequences (namely, HSSKLQ and SSKYQ, both capped with a morpholine ring) that are selectively and efficiently hydrolyzed by the prostate-specific antigen (PSA). These prodrugs showed a half-life of 22 and 3 h, respectively, in the presence of PSA, and strongly affected DU145 prostate cancer cell growth at both 5 and 10 μM concentrations, by reducing cell viability to less than 10% [14].

However, cyclopamine derivatives that maintain the original skeleton are all acid labile compounds. In an attempt to improve stability of cyclopamine derivatives to acidic condition and to ameliorate Hh inhibitory activity, D-ring homologation and oxidation of the alcohol group at C3 were set up at Infinity Pharmaceuticals (Fig. 1) [15]. Starting from natural cyclopamine extracted using previously reported procedures, semisynthetic D-ring homocyclopamine analogues were obtained by C12–C13 stereoselective cyclopropanation followed by acid-catalyzed ring expansion, while an Oppenauer oxidation allowed to obtain a 4-en-3-one system at the A-ring. The new compounds were assayed for their ability to inhibit differentiation of C3H10T1/2 cells to osteoblasts, by monitoring production of alkaline phosphatase (AP). The best compound in terms of both in vitro activity and pharmacokinetic properties was IPI-269609 (**5**) that showed an activity comparable to that of cyclopamine ($IC_{50} = 200$ nM vs 170 nM, respectively), but a 20-fold improved aqueous solubility and a remarkable stability (98%) in simulated gastric fluid. A 3.2 h half-life and an 80% oral bioavailability render such a compound an interesting starting point for further studies on homocyclopamine derivatives [16]. However, the corresponding C3 saturated alcohol and glucuronide conjugate were found as metabolites readily produced after oral administration. As

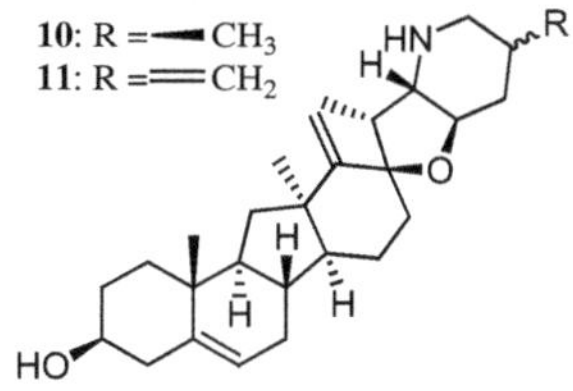

Fig. 2 *Exo*-cyclopamine derivatives (Universität Leipzig and Rice University)

a consequence, IPI-269609 was submitted to a next round optimization by reduction of the enone system to the corresponding *cis*-decalone (**6**), that showed an activity value (13 nM) about 15-fold better than that of the parent compound and a lower metabolic stability in human liver microsomes (about 75 min as half-life). Three different series of compounds were then designed and synthesized starting from the *cis*-decalone scaffold: compounds bearing various groups at C3, compounds bearing heterocycles fused to the A-ring, and compounds with a lactam moiety within the A-ring [17]. Previous structure–activity relationship (SAR) considerations about the C3 positions led to the suggestion that a polar group that contains a hydrogen bond donor and that cannot be transformed into an alcohol group by epatic metabolism should be preferred. In agreement, the best compound of the first series was the C3-sulfonamide (also referred to as IPI-926, saridegib, **7**), characterized by a 7 nM activity and significant metabolic stability (80 min as half-life). Moreover, introduction of a heterocycle fused to the A-ring was also profitable for improving metabolic stability. Both the nature of the heterocycle and the position of heteroatoms were very important for activity. The 2,3-fused pyrazole analogue **8** showed a 13 nM activity and a 120 min half-life. Finally, introduction of a lactam moiety into a homologated seven-membered A-ring (**9**) led to a 25 nM activity and a 65 min half-life. Pharmacokinetic properties of these compounds were good enough to proceed toward assays in animal models of medulloblastoma. The sulfonamido derivative IPI-926 was superior in comparison with the other compounds. In fact, in addition to complete tumor regression, no recurrence occurred during a 21-day post-treatment phase. The molecular target of IPI-926 was identified to be Smo, as demonstrated by its competition for BODIPY-cyclopamine. Recently, a resistant mechanism to IPI-926 was described, based on induction of P-glycoprotein and drug efflux [18]. Studies on IPI-926 have been discontinued.

Recent literature reported further efforts to improve stability and potency of compounds bearing a cyclopamine-like scaffold. A series of cyclopamine derivatives bearing one or more exocyclic double bonds between C13–C18 (D-ring) and C25-C27 (F-ring) were synthesized and tested in a Gli-Luc assay (Fig. 2) [19]. As a result, C13–C18 *exo*-cyclopamine (**10**) and C13–C18/C25–C27 bis-*exo*-cyclopamine (**11**) showed a significantly better (10–20-fold) activity in comparison with that of cyclopamine and were stable at pH 1.

In an attempt to obtain structurally simplified and metabolically stable cyclopamine-like compounds, a bis-aromatic estrone derivative (**12**, Fig. 3) was easily synthesized from estrone and tested for its ability to inhibit Shh-induced signaling activity in a luciferase-based assay, as well as proliferation of mouse

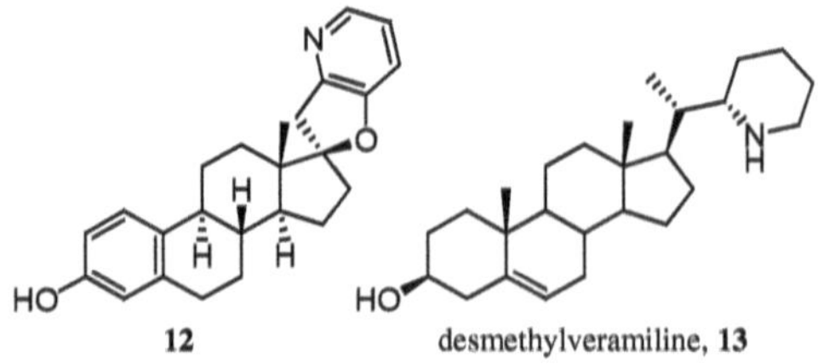

Fig. 3 A bis-aromatic estrone derivative (**12**) and desmethylveramiline (**13**)

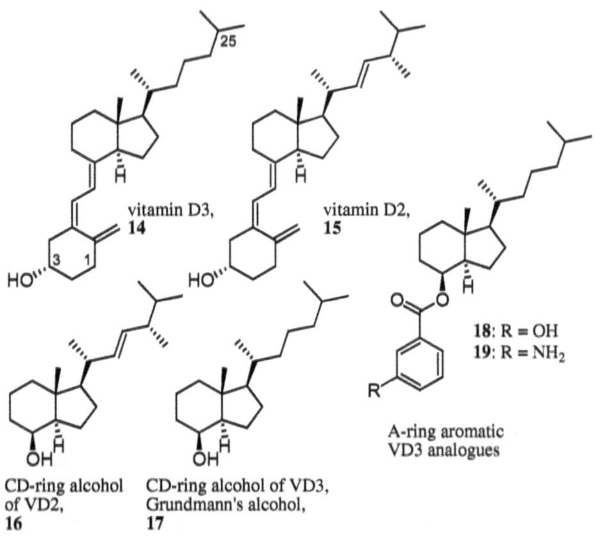

Fig. 4 Vitamin D3 (**14**), vitamin D2 (**15**), their C-D alcohol (**16–17**) and A-ring aromatic derivatives (**18–19**) discovered at the University of Connecticut

granule neuron precursors. The new compound was found to be equipotent with cyclopamine in both assays at a 10 μM concentration [20]. Following a similar approach, desmethylveramiline (**13**, Fig. 3) was synthesized from the commercially available Fernholtz acid and found to be four-fold less active than cyclopamine in a Shh-light2 assay (IC_{50} of 1.1 and 0.3 μM, respectively) [21]. Weak inhibitors of the Hh signaling pathway were also alkaloids contained in common foods, such as solanidine, solasodine, diosgenin, and tomatidine [22]. (although the last was described as inactive toward the Hh pathway by most authors).

2.2 *Vitamin D2 and D3 Analogues (University of Connecticut)*

Vitamin D3 (VD3, **14**, Fig. 4) has been described as an inhibitor of the Hh signaling able to directly target Smo and compete with cyclopamine [23]. In a similar way, also calcitriol (1α,25-dihydroxy derivative of VD3) showed ability to regulate the same pathway [24]. However, enzymes belonging to the Cyp450 cytochrome family metabolize VD3 into the corresponding 25-hydroxy and 1α,25-dihydroxy derivatives that in turn activate vitamin D receptor (VDR). The resulting effects

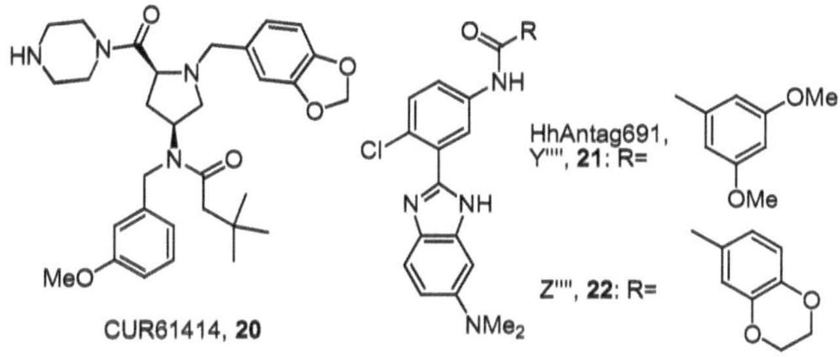

Fig. 5 The pyrrolidine analogue CUR61414 (**20**) and the benzimidazole HhAntag691 (**21**) and Z'''' (**22**) discovered at Curis

comprise an important hypercalcemia. On this basis, Hh modulators with a VD3-like structure should lack metabolism sites in order to avoid VDR activation. Accordingly, various chemical modifications of the A-ring of VD3 were focused at the 3-(*S*)-hydroxyl group. Oxidization to a keto group, methylation, and conversion to amide led to compounds with an activity similar or lower (25–70%) than that of VD3 (32%), measured as percent Gli1 mRNA expression in oxysterol-stimulated C3H10T1/2 cells (relative to control, set to 100%) [25]. Moreover, two of the most active derivatives corresponded to the CD-ring alcohols of vitamin D2 (VD2, **15**) and VD3 (46 and 32% activity) [26]. These compounds (**16** and **17**) were modified by insertion of a phenyl ring reminiscent of the VD3 A-ring linked by an ester moiety. Activity of the resulting compounds, bearing small hydrogen bond donor groups (such as a hydroxyl and amino substituent as in **18** and **19**) mimicking the 3-OH group of VD3, was significantly enhanced (1–2%) in comparison with that of VD3 and its alcohol derivative **17** [27]. Inhibition of Hh signaling in C3H10T1/2 cells was in the micromolar range (IC_{50} values for Gli1 and Ptch spanned from 0.7 to 4.5 μM), binding to VDR was negligible, although Cyp24A1 was upregulated by an unknown mechanism. Additional analogues bearing one or more substituents at the aromatic A-ring, or bearing a heteroaryl or biaryl moiety instead of the A-ring showed Hh inhibitory activity lower than that found for **18** and **19** [28]. Interestingly, these compounds did not bind the cyclopamine binding site on Smo (HEK293T cells), in direct contrast with previous results on VD3 [23]. In particular, anti-Hh activity of VD3 seems to be dependent on VDR, while the molecular target of **18** and **19** still remains unknown. However, a new hypothesis does arise that takes into account the sterol binding site located within the cysteine-rich domain (CRD) of Smo (see Sect. 6.4).

2.3 Pyrrolidines, Benzimidazoles, Biarylamides, and Quinazolinones (Curis–Evotec–Genentech–Roche)

The application of the Gli-Luc assay in a high-throughput screening aimed at finding compounds able to modulate the Hh pathway led to the identification of a pyrrolidine derivative (CUR61414, **20**, Fig. 5) among about 100,000 small

Table 1 Benzimidazoles and pyridine derivatives identified by Curis–Evotec–Genentech–Roche, leading to the identification of GDC-0449 (**27**, vismodegib)

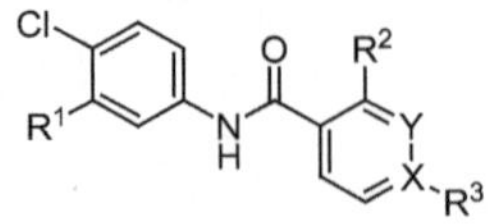

Comp	X	Y	R^1	R^2	R^3	Cell-based assay (IC_{50}, nM)
23	CH	N	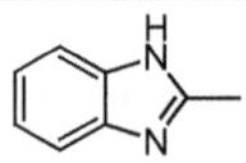	Me	CF_3	12
24	CH	N	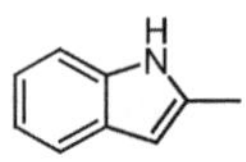	Me	CF_3	500
25	CH	N	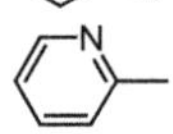	Me	CF_3	42
26	CH	N	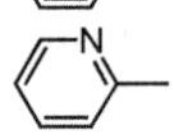	H	H	600
27, *GDC-0449 vismodegib*	CH	CH	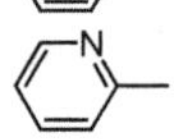	Cl	4-SO_2Me	13
28	CH	CH	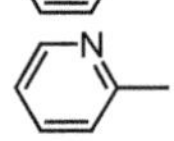	H	H	800
29	CH	N	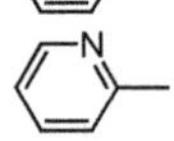	Me	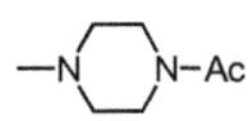	5
30	CH	CH	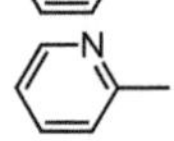	Cl	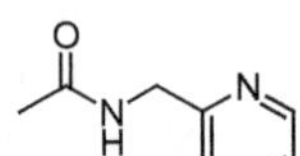	0.4
31	CH	N	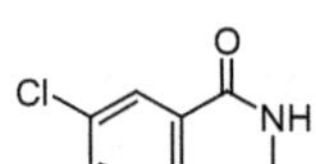	H	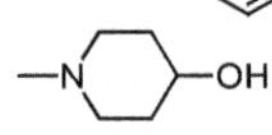	
32	CH	N	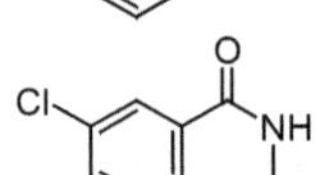	H	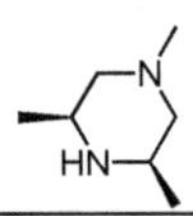	

molecules [29]. Such a compound showed a 100–200 nM IC_{50}, was demonstrated to bind directly to Smo, and was also able to block proliferation of basal cells in carcinoma-like lesions [30]. On this basis, a topical formulation of CUR61414 significantly inhibited skin Hh signaling, blocked the induction of hair follicle anagen, and reduced existing basal cell carcinoma (BCC). Unfortunately, the same formulation gave unsatisfactory results in a phase 1 clinical study on human superficial or nodular BCC [31].

In a subsequent effort made at Curis, a large class of benzimidazole derivatives has been also discovered [32] that included the known Smo antagonist referred to as Hedgehog Antagonist 691 (Hh-Antag691, **21**, labeled as Y'''' in the original patent). Among them, one of the most active compounds (**23**, Table 1) showed a 12 nM activity in the functional Gli-Luc assay on murine embryonic fibroblast cells. This compound, however, suffered from typical weakness among drug candidates: poor metabolic stability, high clearance, and low solubility in water. In a first round optimization [33, 34] a hydrogen bond acceptor group corresponding to one of the nitrogen atoms of the benzimidazole moiety, emerged as a key structural feature for activity. As an example, replacement of the benzimidazole of **23** with an indole ring led to **24** with a 40-fold lower activity (500 nM). Simplification of the indole ring to a pyridine was tolerated only whether the substitution pattern on the ring maintained the nitrogen atom at the optimal position required for the hydrogen bond acceptor (2-pyridyl substituent as in **25**). Activity was maintained at a double-digit nanomolar value (42 nM) and a superior pharmacokinetic profile was found in comparison with **23**. In particular, its clearance in dog was 1.9 mL/min/kg (versus 124 measured for **23**) and solubility was from 3 to 6-fold higher than that of **23**, depending on the pH value (1,000 versus 300 μg/mL at pH 1.0, and 1.8 versus 0.3 μg/mL at pH 6.5). When dosed in vivo, this compound showed, however, poor activity, likely due to dose-limited solubility that in turn resulted in low absorption. To improve pharmacokinetic and dynamics parameters of such a class of compounds, keeping fixed the structure of **25** as a template, structural changes at the benzamide portion were planned. As a result, a small lipophilic group (such as a Me or Cl) at position 2 (R^2) appeared to be important (compare **25** and **27** versus **26**), probably because of its ability to force the terminal aromatic ring and the amide bond to reside in different planes. On the contrary, replacement of the pyridine with a phenyl ring did not influence activity (compare **26** and **28**). A variety of polar substituents added at the para position of the pyridine ring (or the corresponding phenyl ring) were tolerated and resulted in compounds with improved solubility [35]. The most representative example was **27** (GDC-0449, vismodegib) that although a predicted increase in logP showed an unexpected 20-fold improved solubility (>3,000 μg/mL at pH 1.0, and 9.5 at pH 6.5), that is probably consequent to a reduced planarity of the aryl amide system induced by the chlorine substituent. In addition, **27** also exhibited a 13 nM activity in the functional cell-based assay and good metabolic stability. Its low clearance (0.4 mL/min/kg) and high absorption led to high oral bioavailability in dog. Next studies showed oral absorption also in human, in part limited by sub-optimal solubility and pharmacokinetic parameters suitable for clinical trials. The safety and efficacy of GDC-0449 was first demonstrated in patients with advanced BCC with about 55% overall response rate. In addition, treatment of a patient with widespread metastatic medulloblastoma resulted in rapid and dramatic tumor regression [36]. Based on the results of the trials, at the beginning of 2012, GDC-0449 was approved by Food and Drug Administration (FDA) for treatment of locally advanced form of BCC. GDC-0449 is currently under phase 1–2 clinical trials for the treatment of various types (advanced, metastatic, recurrent, and refractory) of solid tumors.

A D473H mutation not detected in primary tumors (before treatment with GDC-0449) led to acquired resistance to GDC-0449 and relapse in both medulloblastoma [36, 37] and BCC [38]. To understand what did occur in consequence of this mutation, mouse C3H10T1/2 cells encoding D477H-mutated Smo (corresponding to the human D473H mutation) demonstrated full ability to activate Hh signaling, as the corresponding wild-type cells are able to do [39]. However, mutated cells are resistant to GDC-0449 and not sensitive to a treatment with KAAD-cyclopamine (IC_{50} value about 45-fold higher than that found in wild-type cells), thus suggesting a reduced ability of Smo to bind these compounds. This hypothesis was then confirmed with radiolabeled GDC-0449 that was unable to bind to mutated Smo, while an expected binding to wild-type Smo was detected. This result was further confirmed in allograft models of medulloblastoma in mice where a mutation was found for the aspartic acid 477 that corresponds to the position 473 in the human sequence. In addition, by means of alanine scan mutagenesis, E518A was identified as an additional mutation site on Smo, responsible for resistance to GDC-0449 [40]. Mutation-driven resistance acquired by tumor cells upon a treatment with a Smo inhibitor strongly suggested the need to identify next generation inhibitors of Smo able to overcome resistance. With this purpose, replacement of the CF_3 and the sulfone substituents of **25** and **27**, respectively, with a number of polar groups led to compounds with activity comparable or better than that of **27** [35]. As an example, **29** and **30** showed higher potency toward both mouse and human cell lines, as well as excellent in vivo activity in an allograft model.

In an alternative approach [40], 53 compounds, including cyclopamine, Smo antagonist 1 (SANT-1, **112**, Fig. 16), Hh-Antag691, GDC-0449 and several of its derivatives, already known as inhibitors of the wild-type Smo, were also assayed for their activity on the D473H and E518K Smo mutants. Following the usual Gli-Luc assay, structurally different compounds showed variable percent Smo inhibition values at 1 μM concentration. Among known compounds, Hh-Antag691 was very active toward both mutants with a >90% inhibition. SANT-1 showed a quantitative inhibition of the E518K mutant (39% inhibition of D473H Smo), while cyclopamine was partially active (48%) toward D473H and completely inactive toward E518K. Interestingly, a class of bis-amides was populated by two compounds (**31** and **32**, Table 1) with significant inhibition of both Smo mutants (>57%). In particular, **32** showed good pharmacokinetic parameters and a 300 and 700 nM inhibitory activity toward wild-type and D473H mutated Smo, respectively. Such a compound was also able to block tumor growth in a mouse model of medulloblastoma and to reduce its volume during a 11-day treatment. Although these two compounds were considered as potential alternative drugs to treat GDC-0449-resistant tumors, their further biological profiling was not published yet.

During a joint research project between Evotec and Curis, the quinazolinone-based aryl urea **33** (Fig. 6) was discovered to have micromolar activity ($IC_{50} = 1{,}400$ nM) in the Gli-Luc assay [41]. A classical medicinal chemistry approach to improve its activity led to rule out several SAR considerations: the condensed phenyl ring was necessary, as a pyrimidinone derivative was ten-fold less active; elision of the fluorophenyl moiety or replacement with alkyl groups (such as a isopropyl substituent) was detrimental for activity (4,300–10,000 nM); the two methyl groups at R^1 and R^2 were unnecessary because des-methyl

Fig. 6 Quinazolinone-based aryl ureas (**33** and **34**) discovered at Curis–Evotec

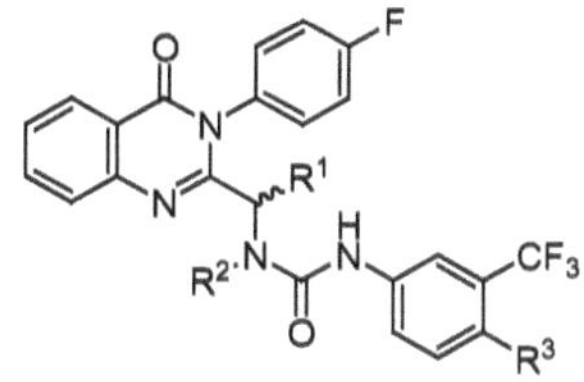

33: $R^1 = R^2 = Me$, $R^3 = H$, $IC_{50} = 1400$ nM
34: $R^1 = R^2 = H$, $R^3 = Cl$, $IC_{50} = 70$ nM

Fig. 7 Benzimidazole SEN450 (35) discovered at Siena Biotech

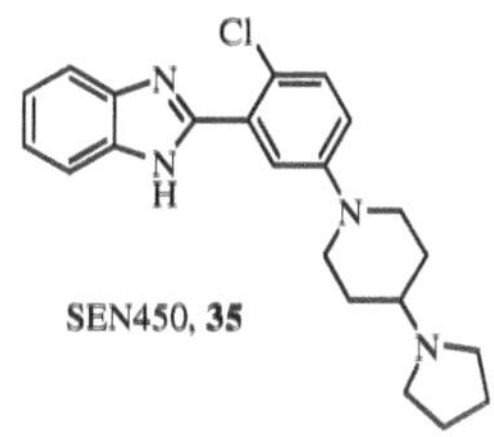

SEN450, **35**

analogues were all more active than the parent compound; further decoration of the trifluoro phenyl moiety of des-methyl derivatives resulted in a significant improvement in activity (70–700 nM) [42]. The best compound **34** showed double-digit nanomolar activity in the functional assay but was not further developed.

2.4 Benzimidazoles (Siena Biotech)

Working on a commercially available recurrent glioblastoma multiforme cell line (DBTRG-05MG), a new benzimidazole derivative (SEN450, **35**, Fig. 7) was identified at Siena Biotech [43, 44]. SEN450 competed with BODIPY-cyclopamine in Flp-In 293 cells and affected alkaline phosphatase activity as a functional inhibitor of Hh signaling in C3H10T1/2 cells [45]. Activity of this compound in both the competition and functional assays was 73 and 23 nM, respectively, similar to that found for GDC-0449 and a HhAntag analogue. SEN450 also showed high solubility, cell permeability, and bioavailability ($F\% = 60$). These results allowed in vivo investigations to evaluate the effect of this compound on tumor growth alone or in combination with temozolomide, the standard chemotherapeutic agent for the treatment of glioblastoma. Although SEN450 showed only moderate activity on tumor growth when administered alone, it was able to block tumor regrowth after temozolomide debulking treatment, without significant side effects (a marginal body weight loss was found). The favorable pharmacokinetic parameters and synergy with temozolomide suggested further studies on SEN450 for setting up combination therapy toward glioblastoma [45].

Table 2 Indazole derivatives disclosed by IRBM-Merck Research Laboratories Rome

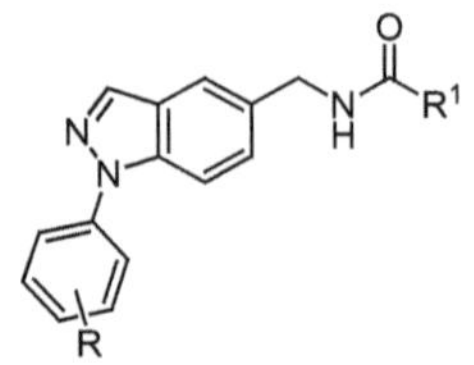

Comp	R	R^1	Smo binding (IC_{50}, nM)	Shh-light2 (IC_{50}, nM)
36	4-F	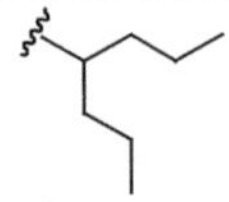	130	150
37	4-F	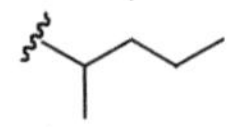	360	230
38	4-F	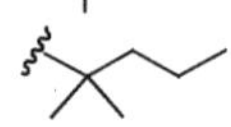	330	150
39	4-F	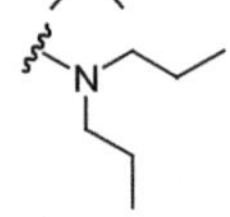	310	490
40	2-OCH_3	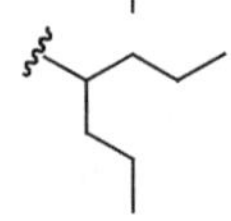	5	5
41	4-OCH_3	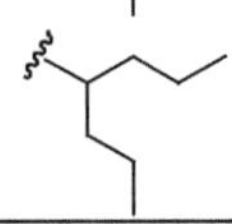	40	10

2.5 *Indazoles, Imidazo-pyridazines, and Oxadiazoles (IRBM-Merck Research Laboratories Rome)*

A high-throughput screening (HTS) campaign was conducted by IRBM-Merck researchers to identify new smoothened antagonists [46]. Compounds were checked for their ability to block the Hh pathway in Shh-light2 cells stably incorporating a Gli-dependent firefly luciferase reporter gene, simultaneously monitoring compound cytotoxicity [11, 47]. Moreover, compounds were also tested in a whole cell assay measuring their ability to compete with BODIPY-cyclopamine and displace it from its binding site on Smo [48]. Indazole **36** (Table 2) was the first compound identified, showing a submicromolar inhibitory activity ($IC_{50} = 150$ nM) without significant cytotoxicity at 5 μM. Its mechanism of action was based on a direct interaction with Smo, as confirmed by its Smo binding affinity ($IC_{50} = 130$ nM). Attempts to optimize activity of **36** showed that most variations of the amide chain (such as shortening, cyclization, and aromatization) resulted in significantly lower binding affinity and activity in the functional assay, with the sole exceptions of α-methyl (**37**) and geminal α,α-dimethyl (**38**) analogues. Also the

Table 3 Tetrahydroimidazo[1,5-*a*]pyrazine disclosed by IRBM-Merck Research Laboratories Rome, leading to the identification of MK-5710

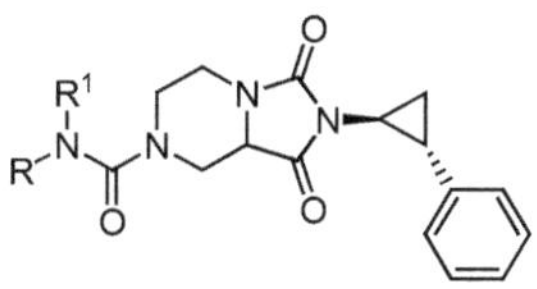

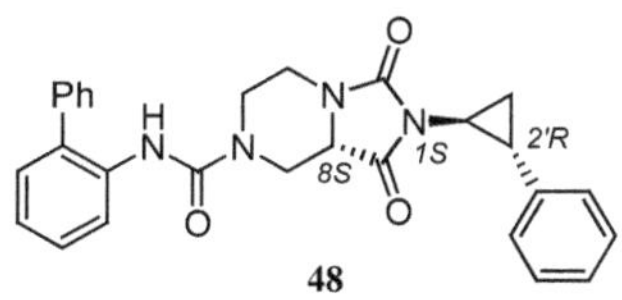

48

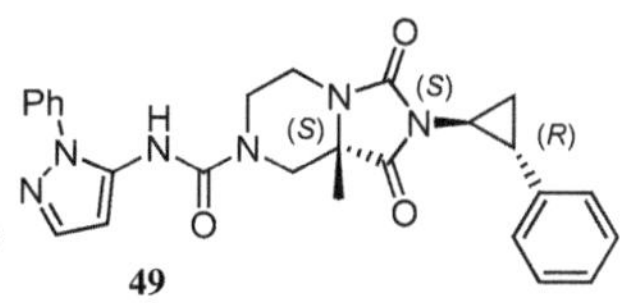

49

Comp	R	R^1	Smo binding (IC_{50}, nM)	Shh-light2 (IC_{50}, nM)
42	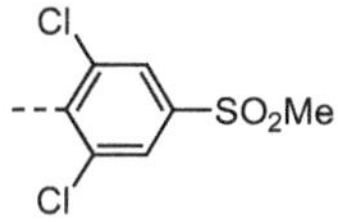	H	77	53
43	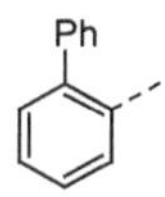	H	94	84
44	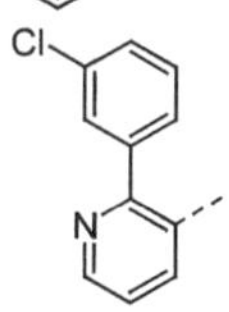	H	19	18
45	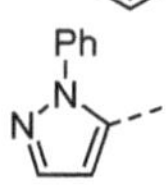	H	13	16
46	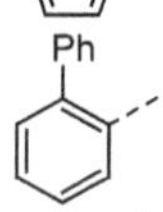	Me	1,000	1,600
47	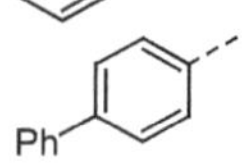	H	890	2,000
48			21	42
49, *MK-5710*			13	17

corresponding urea derivative of **36** (**39**) retained a submicromolar activity. Better results were obtained by decoration of the N1 phenyl ring, with a particular emphasis for ortho (**40**) and para (**41**) methoxy derivatives that showed nanomolar activity.

Another class of potent Smo antagonists was based on the [6,5]-bicyclic tetrahydroimidazo[1,5-*a*]pyrazine-1,3(2*H*,5*H*)-dione scaffold bearing a *trans* 2-phenyl-cyclopropyl moiety at one of the hydantoin nitrogens (Table 3) [49]. Extensive SAR analysis at the left-hand portion of the molecule led to the identification of aromatic substituents as the most profitable for activity. In particular, 2,6-dichlorophenyl derivatives bearing strong electron withdrawing groups at the para position (**42**), as well as biphenyl compounds (**43**) showed very high activity in both Smo binding and functional assays. Additional changes at the biphenyl moiety further improved activity, thus leading to the most potent pyridine (**44**) and pyrazole (**45**) compounds. As general rules, the regiochemistry of cyclopropyl ring was critical for activity. As an example, the *trans* isomer **43** showed an activity in the low nanomolar range (84 and 94 nM in the functional and binding assay, respectively), while the corresponding *cis* isomer only had a micromolar Smo binding affinity. Moreover, the ureidic moiety should have an NH group, being the corresponding N-methylated analogue **46** significantly less active. The substitution pattern on the central ring attached to the urea moiety was critical for activity, being the *para*-substituted biphenyl ring **47** at least one order of magnitude less active than the corresponding *ortho*-substituted analogue **43**. Finally, attempts to modify the bicyclic hydantoin core by aromatization into the corresponding phthalimide, by removal of one of the carbonyl groups, and by ring enlargement to a 6,6-bicyclic scaffold, resulted in a substantial loss in affinity or in completely inactive compounds.

All of these compounds were initially synthesized and tested as mixtures of four stereoisomers, resulting from the stereogenic carbon bridging the hydantoin and piperazine nuclei in combination with racemic *trans* disubstituted cyclopropane ring. On the basis of the nanomolar activity of racemic mixtures, they were deconvoluted by supercritical fluid chromatography (SFC) to give the single stereoisomers that were in turn submitted to the same biological assays [50]. As a result, only slight differences in activity were found among the stereoisomers. Moreover, a chiral synthesis of the biphenyl derivative showed that the most active stereoisomer was characterized by a *1S,8S,2'R* stereochemistry (**48**, Table 3), thus prompting the researchers to use this configuration of the central scaffold to further enlarge SAR considerations. As a result, a series of variously substituted arylpyridines, arylpyrazoles, and phenyl derivatives were found to have a good pharmacokinetic profile with low clearance, excellent oral bioavailability, and moderate half-life. However, a deeper investigation on the possible epimerization of the stereogenic carbon of the hydantoin nucleus showed that a 30% conversion of the phenylpyrazole derivative occurred in both rat and human plasma within 3 h upon treatment. Alkylation of the chiral carbon atom by adding lithium bis (trimethylsilyl)amide (LiHMDS) and methyl iodide (MeI) resulted in methylated diastereoisomers that were in turn separated by chiral SFC. *8S*-derivative **49** showed affinity and functional activity (13 and 17 nM, respectively) comparable to that of the unmethylated analogue (15 and 18 nM, respectively). Moreover, also pharmacokinetic parameters and in vivo activity were maintained. Compound **49**, also referred to as MK-5710 in the current literature, was further profiled in preclinical species (i.e., rat and dog), demonstrating no off-target effects and good pharmacokinetic properties, as well as a significant efficacy in a mouse medulloblastoma allograft model.

The same HTS campaign led to the identification of 4-[3-(quinolin-2-yl)-1,2,4-oxadiazol-5-yl]piperidinyl ureas (Table 4) able to inhibit the Hh pathway and binding of BODIPY-cyclopamine at submicromolar concentrations. Although the modest pharmacokinetic properties, some off-target activity toward human ether-à-go-go-related gene (hERG), and inhibition of cytochrome activity, these compounds were submitted to SAR exploration by structural changes at the quinoline, oxadiazole, and piperidine rings, as well as by introduction of different capping substituents at the

Table 4 Quinolinyl-oxadiazolyl-piperidinyl ureas (IRBM-Merck Research Laboratories Rome)

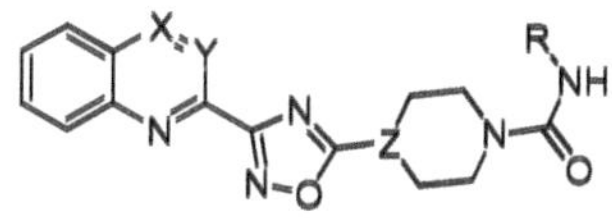

Comp	X	Y	Z	R	Smo binding (IC_{50}, nM)	Shh-light2 (IC_{50}, nM)
50	CH	CH	N	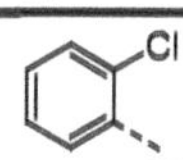	170	140
51	CH	CH	CH	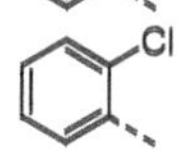	380	510
52	CH	CH	CH	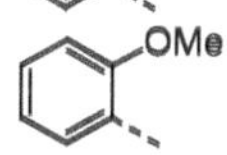	180	410
53	CH	CH	CH	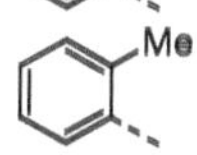	510	440
54	CH	CH	CH	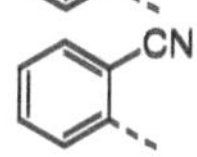	490	190
55	CH	CH	CH	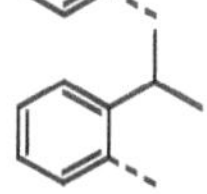	350	480
56	CH	CH	CH	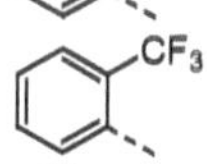	340	700
57	CH	CH	CH	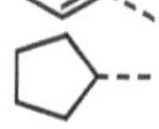	420	630
58	CH	CH	CH	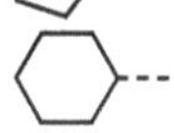	280	57
59	CH	CH	N	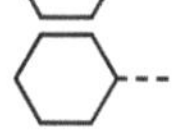	180	53
60	CH	CH	N		39	5
61	NMe	–	N	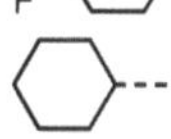	33	5
62	NMe	–	N		25	7
63	CH	CH	CH	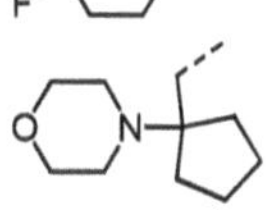	30	46
64	NMe	–	CH	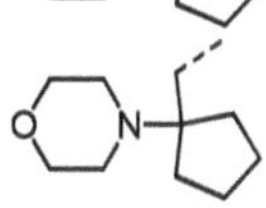	17	18
65	NMe	–	N	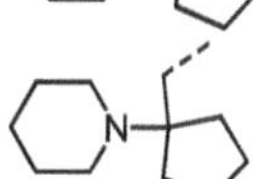	5	4

terminal ureidic moiety. The original 4-[3-(quinolin-2-yl)-1,2,4-oxadiazol-5-yl] piperidinyl scaffold resulted to be the most profitable for activity, only allowing for a replacement of the piperidine ring with an unsubstituted piperazine (**50**) [51]. On the contrary, the ureidic portion of the scaffold was more tolerant to structural variations. In particular, a chloro (**51**), methoxy (**52**), methyl (**53**), cyano (**54**), isopropyl (**55**), and trifluoromethyl (**56**) group at the *ortho*-position of the terminal phenyl ring resulted in compounds with submicromolar activity in Smo binding and Shh-light2 assays. Replacing the terminal phenyl ring with a saturated moiety such as a cyclopentyl (**57**) and cyclohexyl (**58**) group retained or improved activity. Cyclohexyl derivatives showed excellent potency in the range of nanomolar concentration and favorable pharmacokinetic parameters also when the piperidine ring was replaced by a piperazine (**59**, **60**) and when the quinoline was transformed into a benzimidazole (**61**, **62**). Although the introduction of fluorine atoms further improved pharmacokinetic profile in terms of oxidative metabolism and stability in microsomes, such lipophilic compounds suffered from high plasma protein binding. In an attempt to lower plasma protein binding and to improve potency, a diversity screen was performed starting from the cyclopentyl derivative **57** [52]. Excision of the cyclopentyl group to unbranched or branched alkyl spacers, and its homologation to a cyclohexyl or cycloheptyl ring resulted in significantly less active compounds. Interesting results were obtained by introduction of a basic morpholino moiety on the cyclopentyl ring (**63**) that resulted in a inhibition of the Hh pathway more than one order of magnitude better in comparison with **57** (46 vs 630 nM), in addition to a similar improvement of the Smo binding (30 vs 420 nM). More importantly, such a compound also showed a very low plasma protein binding, with a 97% fraction of unbound compound in human plasma. However, the rather lipophilic morpholino derivative was characterized by high intrinsic clearance (higher than 400 μL/min/mg in pooled incubations) determined as turnover in liver microsomes, as well as some off-target effects represented by hERG inhibition ($IC_{50} = 10$ μM). Replacement of the morpholino ring with a more basic amine (such as a pyrrolidine and piperidine) retained a double-digit nanomolar activity, while further improving plasma protein binding and serum shift. Acyclic amines of comparable size were also tolerated. To further enlarge SAR considerations and taking into account previous suggestions showing that the quinoline ring could be profitably replaced by a benzimidazole moiety [51], several derivatives were prepared accordingly and tested [52]. Very interestingly, the analogue of **63** (**64**) showed improved activity and affinity (17 and 18 nM versus 30 and 46 nM, respectively), further ameliorated in the corresponding piperazino derivative **65** (5 and 4 nM).

From a SAR analysis based on the populated series of compounds described above, several considerations could be ruled out. The cyclopentyl group is very important for activity: its opening, homologation, and aromatization led to compounds with lower activity and affinity. The presence of an amino group at the cyclopentyl ring and its basicity affected activity, more basic compounds showing better activity. As expected by previous studies on similar compounds, replacement of the central piperidine with a piperazine, and changing the terminal quinoline fragment with a benzimidazole nucleus positively affected both Hh inhibitory activity and Smo affinity. Lipophilicity is an additional parameter of pivotal importance: increased lipophilicity resulted in higher intrinsic clearance and off-target activity (particularly, hERG inhibitory activity). As a consequence, a

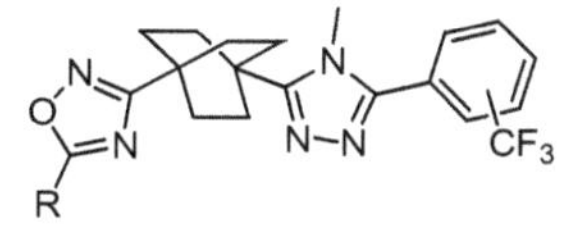
66: R = alkyl or cycloalkyl fluorurated groups

Fig. 8 Oxadiazole derivatives discovered at IRBM-Merck Research Laboratories Rome

reduced lipophilicity was better for activity (since off-target effects were reduced) and for clearance (that was reduced). Finally, although several compounds showed an excellent stability in human liver microsomes, as suggested by a very low intrinsic clearance, the oxadiazole ring was subjected to a reductive ring opening in rats upon iv dosing. This still remains the major limitation of these compounds, since studies to replace the oxadiazole with a metabolically stable moiety are not yet appeared in the literature.

An additional small series of oxadiazoles bearing a terminal phenyltriazole moiety was also disclosed by Merck (**66**, Fig. 8). These compounds were competitors of BODIPY-cyclopamine for Smo. An unspecified triazole showed antiproliferative activity toward murine medulloblastoma cells and induced a significant volume reduction of a medulloblastoma xenograft model at a twice daily dosage of 80 mg/kg [53].

2.6 *Phthalazines, Pyrido-pyridazines, and Bis-amide Derivatives (AMGEN)*

A high-throughput screen performed by AMGEN resulted in a class of phthalazine derivatives with antagonist activity in a functional assay measuring Gli expression in mouse NIH3T3 and human embryonic palatal mesenchymal (HEPM) cells [54, 55]. The parent compound (**67**) showed a 7 nM activity in human cells (Table 5) but suffered from low microsomal stability and high inhibition (47%) of Cyp450 2D6. The first attempt to improve the pharmacokinetic properties of this compound was based on a bioisosteric replacement of the thiophene with a benzene ring, resulting in **68** with a significant reduction of turnover in microsomes and very low cytochrome inhibition (11%). Only a 3-fold decrease in activity toward human cells was observed (19 nM). A SAR study around the left phenyl ring of compound **68** only suggested that simple substituents at the para position (Cl, F, Me) were well tolerated in terms of activity and stability. However, molecular ion mass and mass fragmentation pattern found in mass spectrometry evidenced the piperazine ring as an additional site of metabolism where ring opening, hydroxylation, and sulfatation did occur. As a consequence, a series of methylpiperazine analogues was prepared and tested, among which the 2-(*S*) and the 3-(*R*) methylated derivatives (**69** and **70**, respectively) showed increased potency (5.3 and 2.2 nM) and better microsomal stability. On this basis, a combination of 3-(*R*)-methylpiperazine and small substituents at the para position of the phenyl ring attached to the phthalazine nucleus led to the identification of **71** and **72** with sub-nanomolar activity (IC_{50} 0.7 and

Table 5 Phthalazines, pyrido-pyridazines, and bis-amide derivatives disclosed by AMGEN

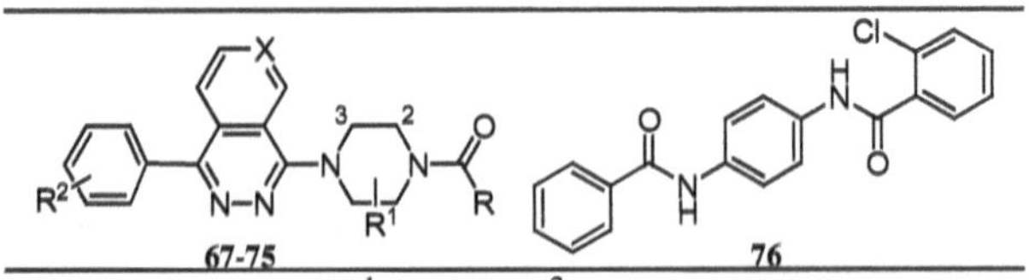

67-75 76

Comp	R	R^1	R^2	X	hSmo (IC_{50}, nM)
67	2-Thienyl	H	H	CH	7.0
68	Ph	H	H	CH	19
69	Ph	2-(*S*)-Me	H	CH	5.3
70	Ph	3-(*R*)-Me	H	CH	2.2
71	Ph	3-(*R*)-Me	4-Cl	CH	0.7
72	Ph	3-(*R*)-Me	4-Me	CH	0.4
73	Ph	3-(*R*)-Me	4-CF_3	CH	2.8
74	Ph	3-(*R*)-Me	H	N	6.0
75	Ph	3-(*R*)-Me	H	N	1.0
76					13

0.4 nM, respectively) and appreciable pharmacokinetic parameters. The trifluoromethyl derivative **73** summarized a nanomolar activity in human cells ($IC_{50} = 2.8$ nM) with low turnover in microsomes and no inhibitory activity toward cytochromes (15%). This compound also showed an excellent in vivo pharmacokinetic profile and significant volume reduction in a medulloblastoma model in mice. The only major limitation of **73** was its pregnane X receptor (PXR) activation liability (81% in comparison with control at a 2 μM concentration), potentially correlated with transactivation of drug metabolizing enzyme, thus resulting in drug–drug interactions. A SAR analysis conducted on a large series of phthalazine derivatives led to the suggestion that PXR activation did depend from molecular hydrophobicity. To check for this hypothesis, the phthalazine scaffold was transformed into a pyrido-pyridazine core by incorporation of a nitrogen atom into the condensed phenyl ring. The optimal regioisomer corresponded to a [3,4-*d*] condensation pattern, resulting in potent compounds **74** and **75** with a 6 and 1 nM IC_{50} toward human Smo, metabolic stability (turnover in both mouse and human microsomes lower that 10%) and negligible PXR activation (5–10%) [56, 57].

An additional high-throughput screening of an in-house compound library led to the identification of the bis-amide **76** bearing a portion of its chemical structure reminiscent of that already published by Genentech [40]. In a human cell-based assay, the new compound showed significant inhibition of Smo (13 nM), but its unsubstituted benzamide moiety was cleaved by *N*-acyl transferase enzymes [58]. Replacement of the metabolizable amide system with a two-atom spacer [such as $COCH_2$, CH=CH, $CH(OH)CH_2$ groups] or with five- and six-membered (hetero)cycles (such as oxazole, thiazole, oxadiazole, and benzene, pyridine, pyrimidine, respectively) resulted in compounds with comparable activity and improved metabolic stability. These results strongly suggested that the amide moiety served only as a geometrical constraint required for locating other molecular portions to the right position to interact with the receptor.

Table 6 Phthalazines and pyridazines identified by Novartis, leading to the identification of NVP-LEQ506

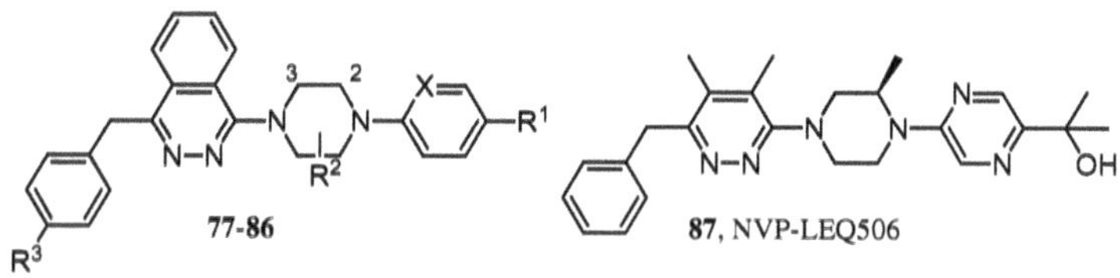

77-86 **87**, NVP-LEQ506

Comp	X	R^1	R^2	R^3	Gli shift (IC_{50}, nM)	hSmo binding (IC_{50}, nM)
77	CH	H	H	H	414	4237
78	N	H	H	H	258	1243
79	N	$C(Me)_2NH_2$	H	H	1.3	15
80	N	$C(Me)_2OH$	H	H	2.7	8
81	N	CH(Me)OH	H	H	38	67
82	N	CH_2OH	H	H	374	484
83	N	CN	2-(*R*)-Me	H	30	34
84	N	CN	3-(*S*)-Me	H	25	17
85	N	CN	H	4-Cl	48	25
86	N	CN	H	4-CN	20	34
87, *NVP-LEQ506*					1	2

2.7 *Phthalazines, Pyridazines, Biaryl Amides, Isoquinolines, and Dihydro-indenes (Novartis)*

A high-throughput cell-based screen of a Novartis in-house database resulted in the identification of a phthalazine derivative hit compound (**77**) with micromolar ability to inhibit the Hh pathway (Table 6) [59]. The unsubstituted benzyl, phthalazine, piperazine, and phenyl moieties of the parent compound **77** were decorated to investigate SAR around each of these rings. To check for their antagonism toward Smo, the resulting compounds were evaluated in the Gli-shift assay and for their ability to displace a radiolabeled known Smo antagonist. In the first assay, two different concentrations of the Hedgehog Agonist 1.5 (Hh-Ag1.5, **197**, Table 18), a known agonist of Smo, were administered to activate Hh pathway. At the highest concentration of Hh-Ag1.5, a Smo competitive inhibitor that directly interacts with Smo is expected to have an activity shift to a higher IC_{50} value. In the second assay, direct binding to Smo was checked by monitoring competition of a compound for the BODIPY-cyclopamine binding site on Smo. An analysis of the SAR suggested that a pyridine moiety (**78**) instead of the terminal phenyl ring was better for activity. Moreover, electron-donating (**79–82**) and bulky (compare **80** versus **81** and **82**) substituents were preferred at the para position of the same ring. The sole structural change allowed at the piperazine ring was the introduction of a methyl substituent, as already found for AMGEN phthalazine derivatives [54, 55]. The resulting four chiral compounds showed a double-digit nanomolar activity toward human Smo. It is noteworthy that the most active methylpiperazine derivatives **83** and **84** showed opposite stereochemistry in comparison with chiral methylpiperazines described by AMGEN [54–57]. Replacement of the phthalazine core with isoquinolines and various 5–6 heterobicycles did not afford better activity. Finally, decoration of the benzyl aromatic portion with small electron-drawing substituents (in particular, Cl and CN) at the para position (as in **85** and **86**) provided a 25 and 34 nM activity, respectively, toward human Smo, comparable to that of the chiral compounds. Among all the phthalazine derivatives, **80** was chosen for further biological profiling, while **79**, that also showed a nanomolar activity, was discarded because of its basic amine terminus that could provide inhibitory properties toward hERG. In vivo pharmacokinetic studies of **80** in mice resulted in encouraging profile: rapid absorption and good oral bioavailability (73%), low plasma clearance, and volume of distribution. Treatment of a medulloblastoma allograft model with **80** (40 mg/kg twice a day) showed an effective dose-related antitumor activity mediated by inhibition of the Hh pathway (Gli1 mRNA inhibition correlated with tumor growth inhibition). This compound, however, suffered from a solubility lower than 5 μM and a significant hERG inhibition ($IC_{50} = 1.5$ μM). In an attempt to increase aqueous solubility and decrease hERG inhibitory activity, opening of the phthalazine core to a pyridazine ring was planned. Two methyl groups at positions 4 and 5 in place of the condensed phenyl ring appeared to be the most profitable substituents for activity. Moreover, a methylated (*R*)-configured chiral center on the piperazine suggested by previous results [59], in addition to a pyrazine ring instead of the pyridine, resulted in **87** (NVP-LEQ506) [60]. Such a compound showed the best combination of activity in the Gli-shift functional assay, low hERG inhibitory activity (30 μM), and acceptable aqueous solubility (34 μM). A direct interaction with Smo was also demonstrated by binding assay, as well as inhibition (96 nM) of C3H10T1/2 cells bearing Smo mutation (D473H) responsible for resistance to GDC-0449 in medulloblastoma patients. The overall efficacy in mouse medulloblastoma allograft model was comparable to that found for sonidegib (NVP-LDE225), suggesting **87** as a very interesting new investigational drug suitable for clinical development.

NVP-LDE225 (**92**, LDE225, sonidegib, erismodegib) was discovered by means of a HTS cell-based screen program on a combinatorial library comprised of 10,000 compounds built by solid-phase synthesis [61]. The parent compound identified by this screening was the biaryl amide **88** bearing heterocyclic moieties at its terminal edges (Table 7). It was submitted to a systematic variation of the two terminal portions, as well as the central phenyl core. Initial SAR speculations suggested insertion of a 4-methyl group on the central phenyl ring and replacement of the benzodioxole moiety with a simpler *p*-cyano phenyl substituent (**89**). Moreover, to avoid metabolic activation possibly occurring at the 1,4-diaminophenyl moiety, the aromatic portion was changed into a pyridine (**90**), while an improvement in activity was found upon introduction of a *cis*-2,6-dimethyl substitution on the morpholine ring (**91**). However, migration of the methyl group from position 4 to 2 on the central ring was the most important structural change leading to compounds with significantly improved activity. The optimal result was obtained by replacement of the *p*-cyano group with a trifluoro methoxy substituent chosen

Table 7 Biaryl amides identified by Novartis, leading to the identification of NVP-LDE225

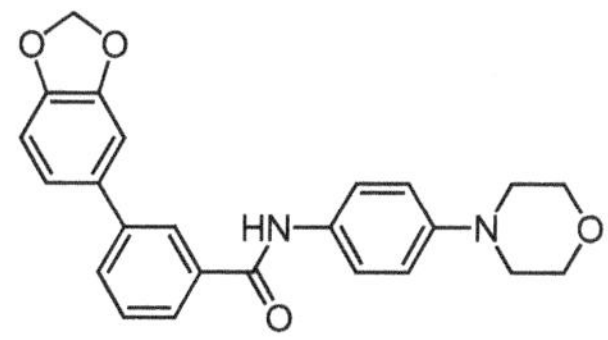

88

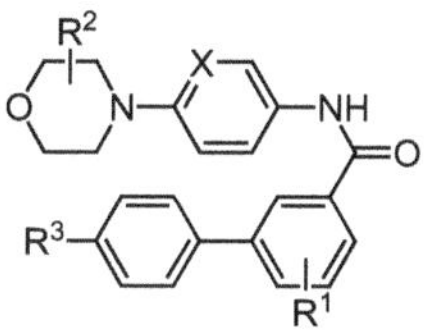

89-92

Comp	X	R^1	R^2	R^3	Gli shift (IC_{50}, nM)	hSmo binding (IC_{50}, nM)
89	CH	4-Me	H	CN	43	
90	N	4-Me	H	CN	4	124
91	N	4-Me	*cis*-2,6-dimethyl	CN	5	12
92, *NVP-LDE225*	N	2-Me	*cis*-2,6-dimethyl	OCF_3	0.6	2.5

among a series of small groups (OMe, CF_3, OCF_3), to afford NVP-LDE225 (**92**). A poor aqueous solubility of the free base prompted chemists to develop a diphosphate salt that increased oral bioavailability to 48% in suspension formulation. LDE225 showed favorable in vitro absorption, distribution, metabolism, excretion, and toxicity (ADMETox) properties and in vivo pharmacokinetic parameters. Treatment with LDE225 did not affect activity of human cytochrome enzymes, and significant off-target effects were not observed. A medulloblastoma allograft model showed tumor regression upon treatment with LDE225, also suggesting the ability of this compound to penetrate the blood–brain barrier in mouse. Unfortunately, a murine model of medulloblastoma treated with LDE225 showed development of resistance, with mutations occurring at five different Smo residues, at least [62].

LDE225 is currently under phase 1–3 clinical trials for the treatment of various types of solid tumors and chronic myelogenous leukemia (CML).

Through a classical medicinal chemistry approach of hit optimization, the known Smo inhibitor Hh-Antag691 (**21**) was submitted to structural changes (Fig. 9) and the resulting compounds were tested for their activity in the Gli-shift assay previously described [59], in the presence of the Hh agonist Hh-Ag1.5 [63]. In particular, the amino-benzimidazole edge was replaced by a phenyl-imidazole group, the benzoyl moiety was rigidified into an isoquinoline, and the metabolically labile dimethoxy groups were omitted and replaced by a morpholino ring, thus leading to **93**. The overall optimization increased activity of Hh-Antag691 by about one order of magnitude (1 vs 0.1 nM activity found for Hh-Antag691 and **93**, respectively). Further changes at the morpholine ring, suggested by previous studies on LDE225, originated the mono- and dimethyl analogues that showed a remarkable systemic exposure (probably consequent to the metabolically labile carbon atoms adjacent to the morpholino oxygen) and moderate plasma clearance. The *cis*-2,6-dimethyl derivative **94** had a 0.1 nM IC_{50} in the Gli-shift assay and in vivo clearance of 247 mL/min/kg.

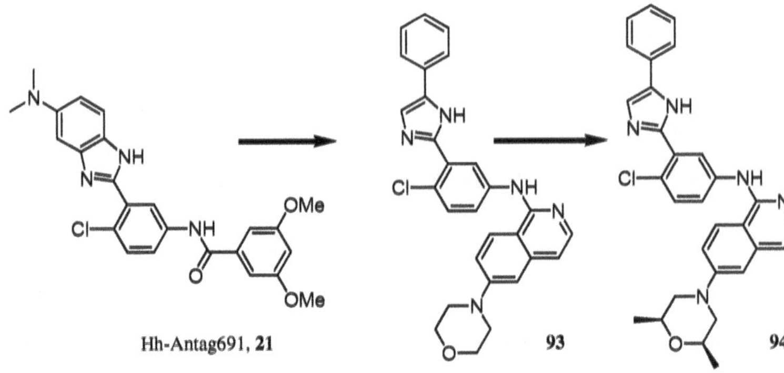

Fig. 9 Rational design of imidazole derivatives **93** and **94** (Novartis compounds) starting from HhAntag691 (**21**)

Table 8 Dihydro-indene derivatives (Novartis)

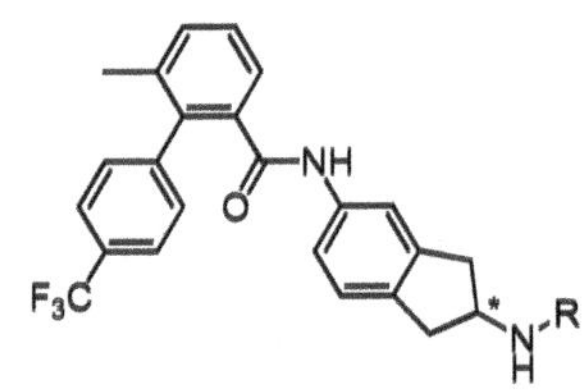

Comp	R	Gli-Luc assay (IC_{50}, nM)	hSmo binding (IC_{50}, nM)
(*R*)-**95**, *LAB687*	COOMe	1,200	3,420
(*S*)-**96**	CH_2-2-thiazolyl	25	30
(*S*)-**97**	CH_2-2-thienyl	10	7
(*S*)-**98**	CH_2-2-furanyl	23	

A parallel effort to optimize ortho-biphenyl carboxamides such as LAB867 [(*R*)-**95**] resulted in dihydro-indene derivatives **96–98** (Table 8), nanomolar inhibitors of the Hh pathway that were able to bind directly to Smo, that however were not further optimized [64].

2.8 Further Phthalazine Derivatives (Eli Lilly)

In addition to AMGEN and Novartis, phthalazine derivatives able to block Hh signaling pathway were also discovered by Eli Lilly researchers. In particular, LY2940680 (**99**, Fig. 10), derived from the corresponding 4-phenyl analogue previously disclosed [65], is a potent and selective antagonist of Smo, with excellent pharmacokinetic properties in various animal species and oral bioavailability

Fig. 10 The phthalazine derivative LY2940680 disclosed by Eli Lilly

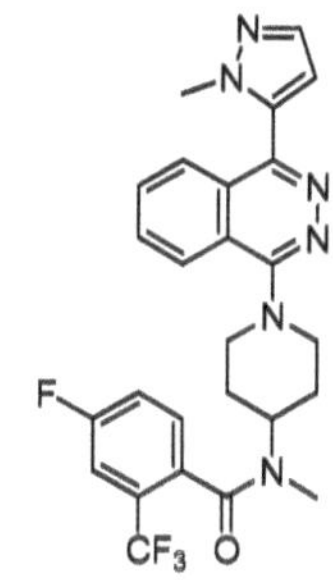

LY2940680, **99**

[66]. Disclosure of the crystal structure of the human Smo receptor bound to LY2940680 [67] confirmed Smo as the molecular target for this small molecule. LY2940680 inhibits Hh signaling in a human medulloblastoma cell line, in C3H10T1/2 mouse cells, as well as in D473H-mutant GDC-0449-resistant cell line. LY2940680 is being investigated in phase 1–2 clinical trials in patients with small-cell lung cancer [68].

2.9 *Quinazolines (Exelixis and Bristol-Myers Squibb)*

A new class of quinazoline derivatives was disclosed by Exelixis and reported to be nanomolar inhibitors of the Hh pathway in a Gli-Luc reporter gene assay. Among hundred of new quinazolines described in the original patent [69], three compounds (**100–102**, Table 9) emerged with IC_{50} lower than 4 nM in Shh-light2 assay. Further studies aimed at developing the quinazoline compounds led to the identification of BMS-833923 (XL-139, **103**). This compound, originally described as an agent for the treatment of ovarian cancer [70], is an inhibitor of Smo currently in phase 1 clinical trial for advanced solid tumors, and in phase 2 in combination with dasatinib in CML patients [71].

2.10 *Imidazo-pyridines (Oslo University)*

Within a focused diversity library of about 12,000 compounds screened by a research group at Oslo University, the imidazo-pyridine MS-0022 (**104**, Fig. 11) was found to compete with BODIPY-cyclopamine and to inhibit Hh pathway in Shh-light2 cells (IC_{50} of 100 nM) [72]. In addition to target Smo, MS-0022 showed an additional activity downstream of SuFu. Similarly to GDC-0449 and differently from cyclopamine, it blocked ciliary accumulation of Smo. At 5–10 μM concentration, MS-0022 inhibited growth of various tumor cell lines and reduced a pancreatic adenocarcinoma xenografts in mice. Attempts to improve activity of this compound by structural changes were unfruitful.

Table 9 Quinazolines identified by Exelixis and Bristol-Myers Squibb, leading to the identification of BMS-833923 (XL-139)

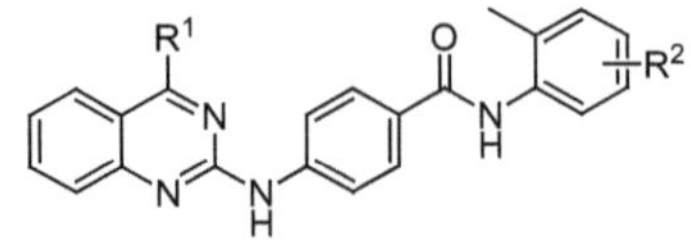

Comp	R[1]	R[2]	Gli-Luc reporter assay (IC_{50}, nM)
100		6-Me	2.8
101	Ph	6-Me	3.6
102	Me	5-$CH_2N(CH_3)CH_2Ph$	3.4
103, *BMS833923, XL-139*[a]	Ph	5-CH_2NHCH_3	

[a]The structure of this compound was described in the original patent (WO 2008 112913, A1)[69] as Example 122, although no activity value was reported at that time

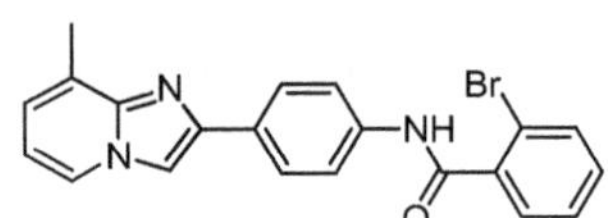

MS-0022, **104**

Fig. 11 Imidazo-pyridine MS-0022 (**104**) discovered at Oslo University

2.11 Thieno-quinolines, Pyrrolo-quinolines, and Pyrrolo-pyridines (Takeda Pharmaceutical Company Limited)

As usually done in recent drug design and discovery approaches, a HTS was applied to identify new scaffold for compounds able to block the Hh pathway. As a result, the thieno[3,2-*c*]quinoline-4-one derivative **105** (Table 10) showed a 5 nM IC_{50} in a Gli-Luc reporter assay, enhanced to 3 nM in the corresponding *N*-methylpyrrole analogue **106** [73]. However, their low metabolic stability precluded any in vivo efficacy. In the attempt to improve the metabolic stability, a SAR investigation was set up by planning changes at the positions 2, 3, and 5 of the scaffold. While the original methoxy group at position 3 and the phenacyl moiety at position 5 were proven as the best substituents for activity, optimization at the position 2 led to significant improvements in terms of metabolic stability in hepatic microsomes. In particular, replacement of the pyrrolidinylethyl side chain of **106** with a 4-piperidinyl ring as in **107** dramatically enhanced metabolic stability (from 95 and 14 μL/min/mg in mouse and human, respectively, to 10 and 2 μL/min/mg), probably because of a significant reduction in lipophilicity of this compound. Moreover, to render the overall length of the molecule comparable to that of **106** without increasing lipophilicity, a hydroxyethyl side chain was added. The resulting compound **108** retained an nM activity in the Gli-Luc assay ($IC_{50} = 6.6$ nM) with a very good clearance in mouse and human (9 and 6 μL/min/mg, respectively). However, a strong ability to inhibit hERG was also found for **108**, probably correlated to its basicity. To avoid this limitation, the hydroxyethyl side chain was modified by introduction of an acyl moiety. Compound **109** retained a potent

Table 10 Thieno-quinoline, pyrrolo-quinoline, and pyrrolo-pyridine compounds disclosed by Takeda Pharmaceutical Company Limited, leading to the identification of TAK-441

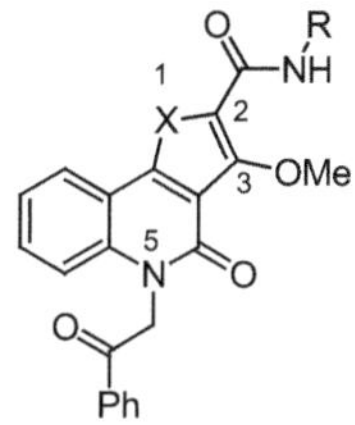

105-109

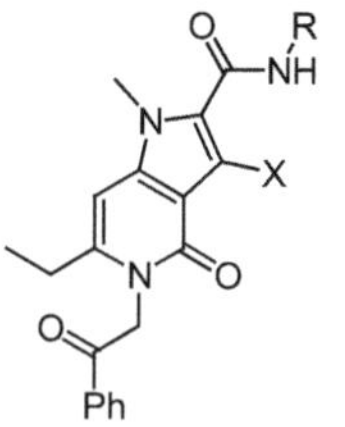

110, 111

Comp	X	R	Gli-Luc reporter assay (IC_{50}, nM)	In vitro clearance (mouse-human) μL/min/mg
105	S	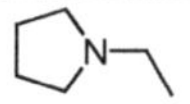	5.1	
106	NMe	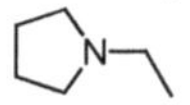	2.9	95-14
107	NMe	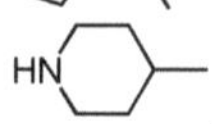	21	10-2
108	NMe	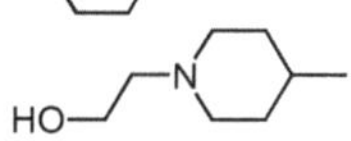	6.6	9-6
109	NMe	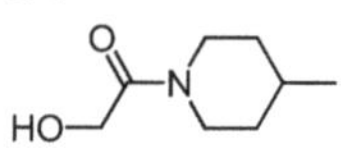	4.6	11-1
110	OMe	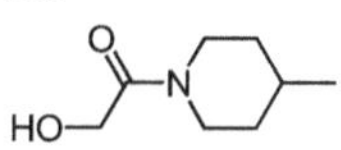	5.7	
111, *TAK-441*	OCH_2CF_3	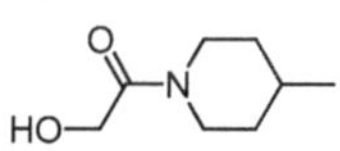	4.4	

activity in the functional assay (4.6 nM) and showed a 40 nM binding affinity toward Smo, without inhibition of hERG. Moreover, in agreement with a very favorable pharmacokinetic profile, such a compound also showed antitumor activity in an allograft model of medulloblastoma. However, further in vivo profiling at high dosage (100 mg/kg) in mice evidenced a poor oral absorption that could be attributed to low solubility (8.4 μg/mL). In the next step of lead optimization, the tricyclic core was simplified by opening the condensed phenyl ring [74]. In particular, the most active (IC_{50} = 5.7 nM) and soluble (63 μg/mL) derivative bearing the simplified bicyclic pyrrolo[3,2-*c*]pyridine core was obtained by insertion of an ethyl group at position 6 of the pyridine ring (**110**). Substituents at the positions 2 and 5 of the original scaffold were kept fixed because they are crucial for activity and metabolic stability. Although a significant improvement of the pharmacokinetic profile, the solubility of **110** was insufficient to guarantee efficacy at higher dose in vivo. The last attempt to improve solubility was based on modification of the alkoxy group at position 3. In fact, while the oxygen atom of this substituent appeared to be necessary for activity, changes in the alkyl portion resulted in increased solubility and enhanced oral absorption. In particular, the study on fluoroethoxy derivatives led to identification of **111** (TAK-441) that showed an 81 μg/mL solubility and a nanomolar inhibitory activity in the functional assay (4.4 nM). TAK-441 was demonstrated to bind directly to Smo at the level of the cyclopamine binding site and was submitted to clinical trials for the treatment of solid tumors [75]. Studies on TAK-441 have been recently discontinued because of project prioritization. Its pharmacokinetic and pharmacodynamics properties were recently used to identify Gli1 mRNA expression in the cancer as a biomarker for predicting the antitumor activity of hedgehog inhibitors [76].

2.12 N-Phenyl-pyrazoles and Benzimidazoles (Pfizer Global Research and Development)

Pfizer researchers pursued a study aimed at identifying Smo antagonists to be used to prevent or slow growth of unwanted hair [77]. This study was based on evidence that Hh pathway is involved in hair cycle regulation and Gli2 activation is necessary for embryonic hair follicle development. To date, the sole eflornithine is used to treat unwanted hair with moderate efficacy. On the contrary, SANT-1 (**112**, Table 11), a previously discovered Smo antagonist [78], showed impressive efficacy in blocking hair growth in a mouse hair growth model. In fact, a preparation containing 3% SANT-1 resulted in significantly higher inhibition of hair growth in comparison with a preparation containing 14% eflornithine. As a consequence, a library screening was undertaken on privileged compounds (*N*-phenyl pyrazoles) bearing structural properties similar to those found in SANT-1. Putative Smo inhibitors were tested on a functional cell-based assay with a β-lactamase transcriptional readout, together with a binding assay to evaluate if Smo was the target of the compounds. Several phenyl-pyrazole were synthesized for SAR purposes, taking into account the structure of SANT-1. The hydrazone moiety of SANT-1 was removed to avoid possible hepatotoxicity and neurotoxicity, resulting in phenyl-pyrazoles bearing a phenyl-piperazine moiety at position 4 of the heterocycle. Keeping fixed the piperazine, the remaining three cycles were variously decorated with small substituents (F, Me, OMe) leading to **113** and **114** as the most active compounds in both the functional and binding assays. Although affinity values were comparable to that of SANT-1, activity in functional cell-based assay was significantly lower. In a next step, to avoid the possible toxicity associated with the presence of an aniline moiety, the piperazine ring was modified by removing its anilino nitrogen and contracting the ring size to five and four atoms. Unfortunately, the resulting compounds **115** and **116** suffered from lower activity toward Smo and low selectivity in a panel of GPCR. However, **116** showed ADME properties desirable for a topical drug: short half-life (10 and 30 min in rat and human liver microsomes, respectively), high membrane, and skin penetration. Moreover, the same compounds did not bind cytochromes and did not inhibit hERG (56% inhibition at 3 μM). Finally, it reduced hair growth in a mouse model, with a lower efficacy in comparison with eflornithine.

Table 11 *N*-phenyl-pyrazoles identified by Pfizer

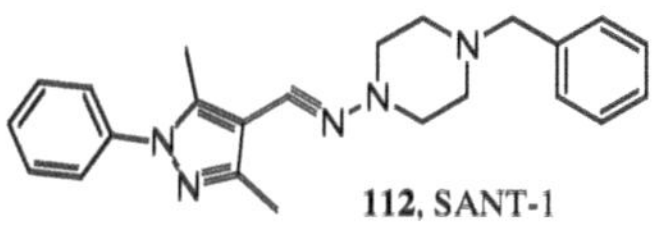

112, SANT-1

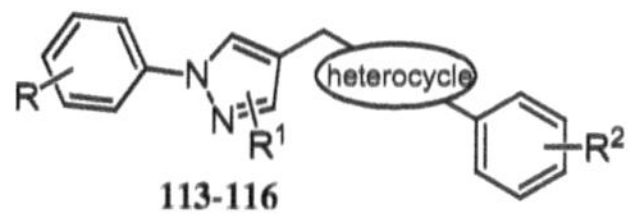

113-116

Comp	Heterocycle	R	R^1	R^2	Cell-based assay (IC_{50}, nM)	Smo binding (IC_{50}, nM)
112, *SANT-1*					5	25
113		4-F	3,5-Dimethyl	3-OMe	823	20
114	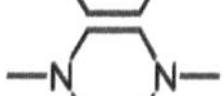	4-F	5-Methyl	3-OMe	65	5
115	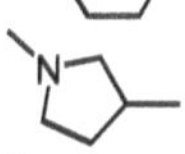	2-F	5-Methyl	2,5-diOMe	41	216
116	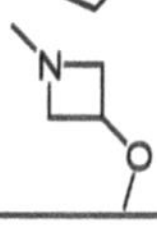	H	5-Methyl	2-F,6-OMe	8	124

Table 12 Benzimidazoles identified by Pfizer, leading to the phenyl-urea derivative PF-04449913

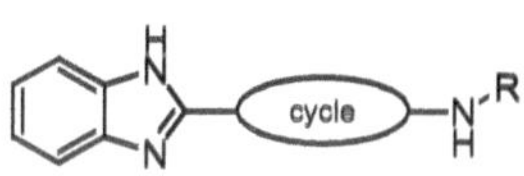

Comp	Cycle	R	Cell-based assay (IC_{50}, nM)
117	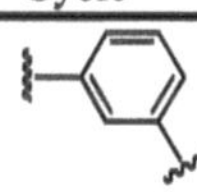	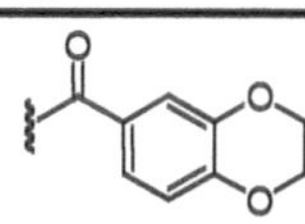	5
118	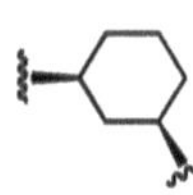	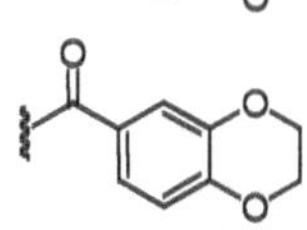	88
119	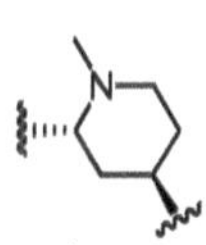	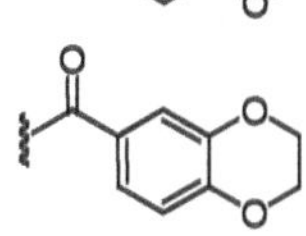	43
120, *PF-04449913*	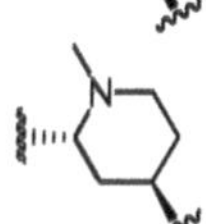	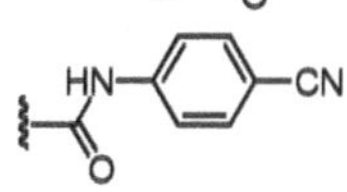	5

Another approach led Pfizer's researchers to identify PF-04449913 (**120**, Table 12) [79], a very potent and orally bioavailable inhibitor of Smo. The parent compound (**117**) for this study belonged to the benzimidazole class previously

Fig. 12 The bipyridinyl derivative PF-5274857 (**121**) identified by Pfizer

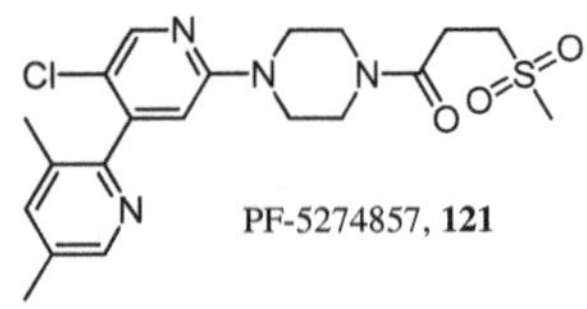

PF-5274857, **121**

discovered at Curis, comprised of many compounds with nanomolar activity in the Gli-Luc reporter assay [32]. Although the excellent potency (**117** had a 5 nM activity in a Gli-Luc assay), such compounds showed low metabolic stability and quantitative binding to plasma proteins due to their high lipophilicity. In the first attempt to reduce lipophilicity and to understand the role played by the central phenyl ring, it was replaced by a cyclohexyl moiety in a 1*R*,3*S* *cis*-1,3-disubstituted configuration. This compound (**118**) showed better metabolic stability, but a 4% oral bioavailability consequent to a poor solubility (37 μg/mL). Since no improvement of solubility was obtained by decoration of the two terminal bicyclic moieties, the central cyclohexyl ring was replaced by a more polar piperidine (**119**) that led to a two-fold enhancement in activity ($IC_{50} = 43$ nM) and improved pharmacokinetic parameters. Metabolic stability was maintained, while oral bioavailability was 33%. An extensive SAR study to identify the optimal substituent for position 4 of the piperidine ring led to synthesize a series of phenyl urea analogues, among which the *p*-cyano derivative (**120**, PF-04449913) showed the best combination of properties. About 10% compound was found free in human plasma, without inhibition of cytochromes. On the other hand, solubility was low (0.02 mg/mL) in water but it was high in simulated gastric fluid (70 mg/mL). The overall profile of this compound in terms of potency ($IC_{50} = 5$ nM) and pharmacokinetic properties prompted the researchers to advance it to in vivo studies. A potent dose-dependent inhibition of the Hh pathway activity and a consequent tumor regression was found in mouse models of medulloblastoma upon treatment with PF-04449913 alone or in combination with other chemotherapeutic agents [80]. Mice also showed inhibition of Gli1 gene expression and down regulation of many genes involved in the Hh signaling pathway.

PF-04449913 is currently under phase 1 clinical trials for the treatment of advanced solid tumors and refractory hematologic malignancies, while it is in phase 2 for the treatment of relapsed/refractory acute myeloid leukemia.

Finally, the bipyridinyl derivative PF-5274857 (**121**, Fig. 12) was also described as a potent and selective Smo antagonist that directly binds Smo and completely abrogates Gli1 transcriptional activity [81]. Affinity of this compound for Smo was found to be 4.6 nM (K_i), while its inhibition toward Gli1 activity in a functional assay was 2.7 nM (IC_{50}). PF-5274857 is orally available, penetrates the blood–brain barrier both in medulloblastoma mice and in nontumor-bearing mice, and is metabolically stable in vivo. Given the significant correlation between its pharmacokinetic and pharmacodynamics properties, PF-5274857 was considered as a lead compound suitable for clinical development.

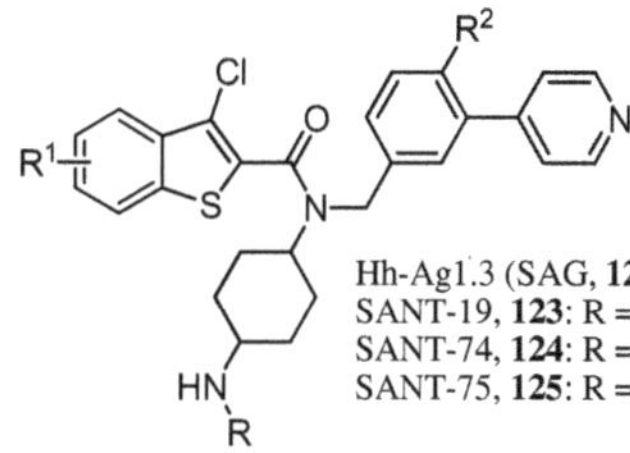

Hh-Ag1.3 (SAG, **122**): R = Me
SANT-19, **123**: R = Et
SANT-74, **124**: R = Allyl
SANT-75, **125**: R = Pr

Fig. 13 Smo antagonists derived from SAG

2.13 *Benzothiophene Derivatives of the Smoothened Agonist (SAG) (Peking University and the University of California)*

A small library (48 compounds, Fig. 13) of SAG (**122**, Table 18) analogues was obtained by parallel synthesis to rule out SAR considerations on benzothiophene derivatives and was assayed in a transgenic zebrafish model [82]. As a result, SANT-19 (**123**) blocked Gli1-green fluorescent protein (GFP) expression in zebrafish embryo at about 80 μM, similarly to the treatment with 20 μM cyclopamine. Surprisingly, SANT-19 and SAG showed a very high structural similarity, the only difference being an N-ethylamino side chain instead of an N-methylamino side chain, respectively. This result strongly suggested that simple variations at the amino side chain of SAG could result in changing from an agonist to an antagonist profile toward Hh pathway. Accordingly, the allyl- and propylamino analogues [SANT-74 (**124**) and SANT-75 (**125**)] identified in a second small library (13 compounds) abolished Gli1-GFP expression at 20 μM or less. Bulkier substituents (butyl and benzyl) were, however, inactive. Treatment of Shh-light2 cells with SANT-19, −74, and −75 showed IC_{50} values better than that found for cyclopamine (200, 70, 20, and 250 nM, respectively). Moreover, SANT-75 also competed with BODIPY-cyclopamine, as also SAG did. Finally, a fluorescence resonance energy transfer (FRET) analysis further supported previous hypothesis [47] that Smo antagonists, such as SANT compounds, could bind to a closed and inactive conformation of Smo, while agonists (such as SAG) prompted Smo to an open and active conformation. Subsequent studies provided evidence of higher complexity in activation of Hh pathway, also depending on Smo translocation to the primary cilium. As an example, it is known that cyclopamine block Smo in an inactive conformation able to migrate to the primary cilium, while SANT-1 could induce formation of a different form of inactive Smo insensitive to cyclopamine stimulation to traffick toward the cilium [83]. A more detailed model of Smo activation is based on subcellular localization of Smo (cytoplasm and primary cilium) and different Smo conformations (active and inactive states): cytoplasmic inactive Smo conformation (able to interact with SANT) moves to the primary cilium in its inactive state (able to interact with cyclopamine and Ptch), and is then activated here by Shh or Smo agonists [84]. The structures of the complexes (protein data bank entries 4n4w and 4o9r) between Smo and two inhibitors (SANT-

1 and cyclopamine) [85] shed light on the interaction pathway among the protein and its ligands. In particular, there is definitive evidence that SANT-1 and cyclopamine are bound to different pockets of Smo (Fig. 17). Moreover, a comparison of the two complexes resulted in a 0.87 Å root-mean square deviation calculated on the basis of a pairwise Cα alignment on the common amino acid sequence (192–548). This clearly means that the structure of Smo around the binding sites that accommodated SANT-1 and cyclopamine is significantly similar in terms of three-dimensional arrangment and did not allow for a differentiation between an active and an inactive Smo conformation.

By a diversity-oriented synthesis (DOS) program, second generation SANT-75 derivatives were synthesized by systematic changes at four structural portions of the molecule (benzothiophene, cyclohexylamine, central benzene, terminal heterocycle), but results from biological investigation are not publicly available yet [86].

2.14 Acylurea, Acylthiourea, and Acylguanidine Derivatives (MRT Compounds) Identified by Virtual Screening (University of Siena, CNRS Gif-sur-Yvette, Université de Strasbourg)

By application of a ligand-based drug discovery approach, a pharmacophoric model has been built starting from known Smo inhibitors. The pharmacophore has been used as a three-dimensional query to filter in silico databases of commercially available compounds. An acylthiourea (MRT-10, **126**, Table 13) was identified that was able to inhibit Shh-light2 cell luciferase activity and SAG-induced C3H10T1/2 cell differentiation in the range of micromolar concentration (IC_{50} of 640 and 900 nM, respectively), as also found for reference compounds cyclopamine and CUR-61414 [87]. MRT-10 competitively inhibit binding of BODIPY-cyclopamine to Smo with an $IC_{50} = 500$ nM. Efforts to identify optimized compounds with higher potency in comparison with MRT-10 led to the synthesis of the corresponding acylurea MRT-14 (**127**) and acylguanidine MRT-83 (**128**) [88]. MRT-14 showed a 4-fold improvement with respect to the parent compound, while activity of MRT-83 was 20–60-fold better than that found for both cyclopamine and MRT-10. MRT-83 also blocked Smo trafficking to the primary cilium and inhibited Hh signaling in vivo (within the subventricular zone of the lateral ventricle of adult mice). SAR analysis suggested that the trimethoxy arrangement is required for activity, the acylguanidino linker is preferred to both acylurea and acylthiourea, a methyl group decorating the central phenyl ring is profitable for activity, and a bulky hydrophobic moiety is required at the remaining edge of the molecule, as demonstrated by MRT-83 and its *p*-F derivative **129** [89].

Table 13 Acylurea, acylthiourea, and acylguanidine derivatives (MRT compounds) identified by virtual screening

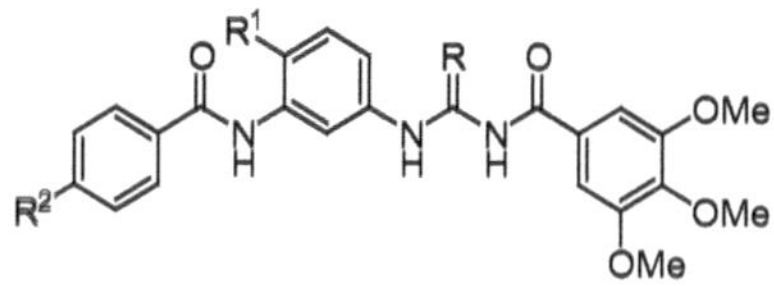

Comp	R	R^1	R^2	Shh-light2 (IC_{50}, nM)	C3H10T1/2 (IC_{50}, nM)	Bodipy-cyclopamine binding (IC_{50}, nM)
126, *MRT-10*	S	H	H	640	900	500
127, *MRT-14*	O	H	H	160	130	120
128, *MRT-83*	NH	Me	Ph	15	10	14 (5)[a]
129	NH	Me	4-F-Ph	14	11	
Cyclopamine				300	620	50

[a]In parentheses, binding toward human cells

2.15 *Hh Signaling Inhibitors Found Among Compounds with Relevant Human Exposure and Environmental Toxicants*

The knowledge that activated Smo binds β-arrestin 2 was used to set up a HTS for the identification of both Smo agonists and antagonists. Co-expression of Smo and GFP-conjugated β-arrestin 2 induced formation of GFP-β-arrestin 2 aggregates that can be identified by visual inspection. On the other hand, the presence of a Smo inhibitor could be easily identified since it prevents formation of intracellular aggregates. An intriguing study that applied this assay was performed to identify compounds with relevant human exposure that are able to affect Hh signaling pathway and possibly contribute to the incidence of the cranio-facial abnormalities known to be caused also by cyclopamine. Among 4,240 natural products, pesticides, and compounds of pharmaceutical interest, ipriflavone (**130**, a dietary supplement, Fig. 14), tolnaftate (**131**, an antifungal drug), and 17-β-estradiol (**132**, a human hormone) were identified [90]. They inhibited Hh signaling in both mouse and human cells with potency 10–30-fold lower than that of cyclopamine. Their molecular target is located between Ptch1 and Gli1 but it was not definitely identified.

In another study, 1,400 environmental toxicants were assayed for their ability to affect Hh pathway [91]. The insecticide synergist piperonyl butoxide (PBO, **133**, Fig. 14) was identified as a compound able to bind Smo and block Smo-induced Gli-Luc reporter activity. It is worth noting that a correlation could exist between defects of mental functions in children exposed to PBO and the inhibition of Hh pathway caused by PBO itself.

Finally, screening a library of about 5,800 commercially available compounds led to the identification of A8 (**134**, Fig. 14), a competitive antagonist of Smo that is

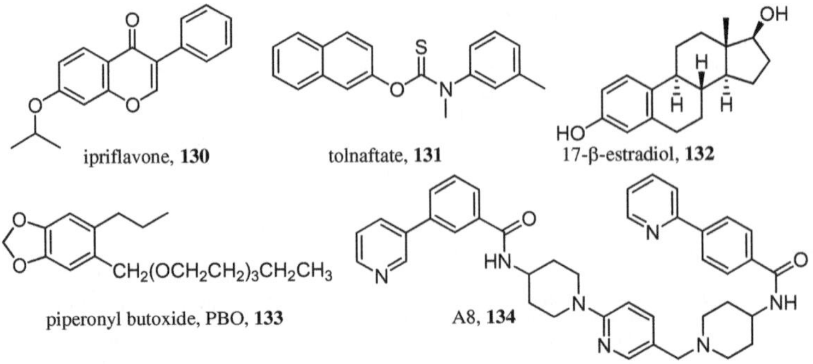

Fig. 14 Pathway inhibitors found among compounds with relevant human exposure and environmental toxicants

able to displace labeled cyclopamine with affinity comparable to that of GDC-0449 and LDE-225 (K_i values were 12, 16, and 38 nM, respectively) [92]. A8 was also able to displace cyclopamine from D473H mutated Smo (associated with clinical resistance in medulloblastoma), differently from GDC-0449 and LDE-225. Moreover, a nanomolar inhibitory activity ($IC_{50} = 2.6$ nM) was also found toward Gli-reporter activity in Shh-light2 cells.

3 Compounds Affecting the Hh Signaling Pathway with Direct Evidence of Smo Binding at Allosteric Sites

3.1 Allosteric Binders of Smo: SANT Compounds (Johns Hopkins University)

Through the Shh-light2 assay in a high-throughput format, four small molecules (**112**, **135–137**, Fig. 15) able to inhibit the Hh pathway were identified by researchers at the Johns Hopkins University [78, 93] and referred to as SANT-1-4 derivatives. These compounds inhibited Hh activity with a 20–200 nM IC_{50} and competed with BODIPY-cyclopamine, thus suggesting their direct interaction with the same binding site on Smo. However, details of experimental assays showed differences in their specific mechanism of action, prompting the authors to hypothesize that different conformations of Smo can exist [93]. Further studies on SANT-1 and SANT-2 suggested allosteric interactions at Smo [94]. Biochemical characterization was performed by evaluation of binding affinity of known Smo agonists and antagonists in the presence of radiolabeled SAG and cyclopamine. Binding of SAG to Smo was fully inhibited by all the test compounds [Hh-Ag1.5, cyclopamine, KAAD-cyclopamine, Z'''' (**22**, Fig. 5), **96**, and GDG-0449], including purmorphamine (**200**, Fig. 25) and tomatidine that were weak inhibitor of SAG binding (inactive up to 10 μM). The only exception was represented by SANT-1

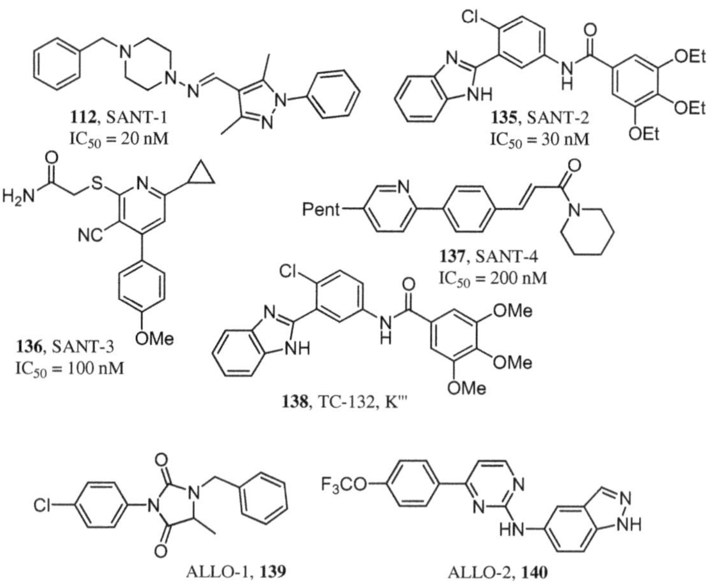

Fig. 15 Allosteric binders of Smo: SANT and ALLO compounds

that showed a partial inhibition of SAG binding, likely consequent to an allosteric interaction of SANT-1 on Smo (further supported by subsequent X-ray crystallography studies, see below). SANT-2 also showed an allosteric profile. On the other hand, when the labeled probe was changed to cyclopamine, full displacement was observed also for SANT compounds, thus suggesting a competitive antagonism. Taken together, these results prompted to hypothesize a probe-dependent allosteric binding of SANT compounds. An additional noteworthy result showed purmorphamine as a weak inhibitor of BODIPY-cyclopamine binding [94] (in agreement with previous results) [95], but unable to displace labeled SAG or cyclopamine. This behavior is similar to that found for ALLO1 (**139**, see below) and could be consequent to an allosteric binding with respect to SAG or cyclopamine. An attempt to find compounds better than SANT-2 by changes at the trialkoxyphenyl and the indolyl moieties led to a series unimportant derivatives [96]. Only one molecule (**138**, referred to as TC132, structurally related to Hh-Antag691 and corresponding to the trimethoxy analogue of SANT-2), already claimed as compound K‴ in the original Curis' patent [32], showed an activity comparable to that of the parent compound (80 vs 98 nM, respectively).

Very recent studies on Smo-targeted Hh antagonists definitively addressed the issue of how cyclopamine, SANT-1, and LY2940680 bind the Smo protein. Superposition of the three-dimensional structures of the Smo-inhibitor complexes (entry 4n4w for SANT-1, 4jkv for LY2940680 [67], and 4o9r for cyclopamine) [85] showed that the inhibitors do not share exactly the same binding site (Fig. 16). In fact, part of the structure of LY2940680 was superposable with part of SANT-1, while the remaining portion of LY2940680 was superposable with cyclopamine. In

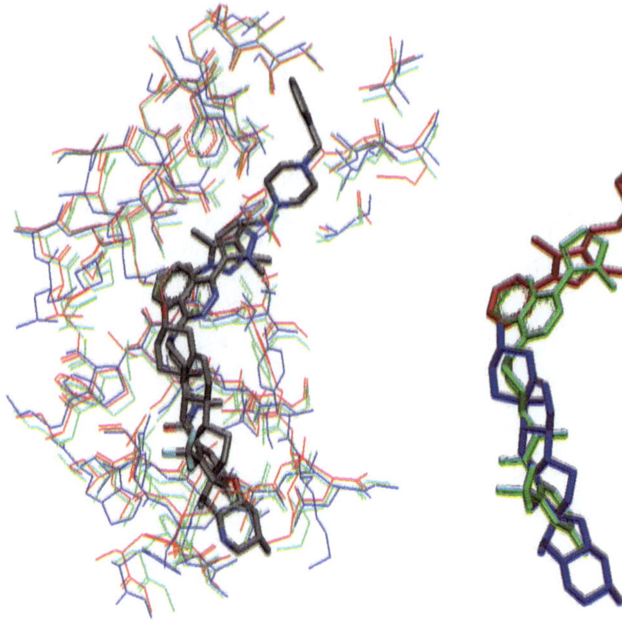

Fig. 16 Graphical representation of the X-ray three-dimensional structure of three known Hh inhibitors that target Smo. *Left*: Superposition of the Smo inhibitor complexes 4n4w (*red*, inhibitor SANT-1), 4jkv (*green*, inhibitor LY2940680), and 4o9r (*blue*, inhibitor cyclopamine) stored in the Protein Data Bank. The inhibitor structures are colored by atom type notation (carbon: black; oxygen: red; nitrogen: blue; fluoride: light blue). For the sake of clarity, only a few amino acids are displayed, without explicit hydrogen atoms. *Right*: Reciprocal orientation of the three Smo inhibitors as extracted by superposed complexes. SANT-1 is *red*, LY2940680 is *green*, and cyclopamine is *blue*. Partial match between SANT-1 and LY2940680, as well as between LY2940680 and cyclopamine is observable. SANT-1 and cyclopamine are accommodated within two different regions of the Smo binding site. Their shortest distance (about 1.3 Å) was found between the 3-hydroxyl oxygen atom of cyclopamine and one of the meta carbon atoms at the N1-phenyl ring of SANT-1

detail, the condensed phenyl ring, the pyrazole moiety, and the N-methyl group of LY2940680 corresponded to the N1 phenyl ring, the pyrazole moiety, and the 3-methyl group of SANT-1, respectively. On the other hand, the piperidine ring, the amide linker, and the terminal phenyl ring of LY2940680 matched the A-ring, the C-ring, and the D-E rings of cyclopamine, respectively. SANT-1 and cyclopamine did not share any common molecular portion, thus interacting with different regions of the Smo protein. The shortest distance (about 1.3 Å) was found between the cyclopamine hydroxyl oxygen (position 3) and one of the meta carbon atoms of the N1-phenyl ring of SANT-1.

Moreover, a certain number of literature reports describe Smo in two different forms, as already found for other G protein-coupled receptors (GPCR): an open active conformer and a closed inactive form. It is known that a conformational shift of the cytoplasmic ends of transmembrane helix 6 (TM6) and TM7 is responsible for the transition from the active to the inactive form of GPCR [11]. In further detail, the closed inactive form of Smo shows the intracellular loop 3 (located between TM5 and TM6) in close proximity to its C terminus [97], and the conformation of TM6 is very close to that of TM2 (less than 8 Å) in the complex between Smo and LY2940680 [67]. Transition from the active to the inactive conformation can be induced by small molecules. As an example, cyclopamine induces Smo to have an inactive conformation [11, 83].

3.2 *Allosteric Binders of Smo: ALLO Compounds (Scripps Research Institute and Genomics Institute of the Novartis Research)*

ALLO1 (**139**) and ALLO2 (**140**, Fig. 15) are Hh signaling inhibitors identified by a HTS of about 50,000 commercially available compounds [98]. They inhibited Gli-Luc activity with nanomolar IC_{50} (50 and 6 nM, respectively) in cells where Hh pathway was activated by Hh-Ag1.5 or ShhN. ALLO1 and ALLO2 did not compete with radiolabeled Hh-Ag1.5, competed with BODIPY-cyclopamine, and had a different profile in the radiolabeled cyclopamine competition assay. In fact, while ALLO1 did not compete with cyclopamine, ALLO2 did. These results, in addition to evidence that ALLO1 and ALLO2 acted upstream of SuFu, led to the suggestion that at least ALLO1 could bind Smo at a binding site different from that of cyclopamine and likely corresponding to the Smo pocket that accommodated the BODIPY moiety. Moreover, their binding site was also different from that of SAG and its Hh-Ag1.5 analogue. Interesting results were obtained by evaluating ALLO activity toward Smo mutants. ALLO compounds inhibited both wild-type and D477G mutated Smo in TM3 cells with a 2-fold IC_{50} shift in comparison with a 185-fold shift found for GDC-0449. Moreover, they also inhibited the corresponding human D473G Smo mutant and showed antiproliferative activity on medulloblastoma cell lines bearing or not a D477G mutated Smo. Finally, the W539L mutation of an oncogenic form of Smo probably induced some changes at the level of the cyclopamine binding site on Smo. In fact, while cyclopamine and ALLO2 (that did compete with cyclopamine) were unable to inhibit W539L Smo-activated Hh pathway, ALLO1 (that did not compete with cyclopamine) showed a significant activity. This could mean that such a mutation disrupted cyclopamine binding site, but did not affect activity of Smo inhibitors non-competitive with cyclopamine and acting at different binding sites.

3.3 *Allosteric Binders of Smo: Smoothened Mutant Antagonists (SMANT) (Harvard University)*

A very interesting approach for the identification of Hh antagonists was pursued at Harvard University by direct analysis of Smo ciliary translocation. This assay was

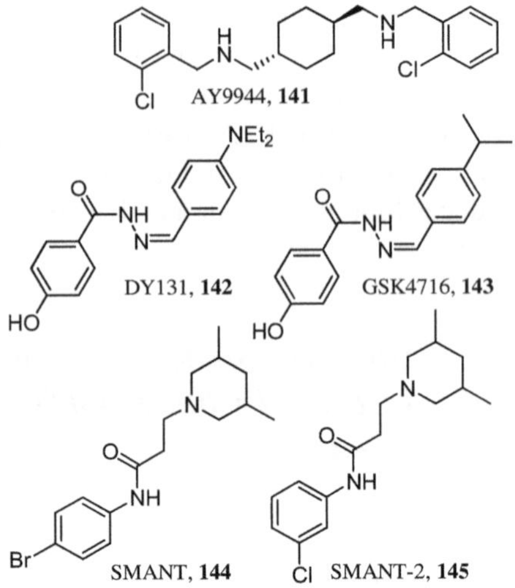

Fig. 17 Allosteric binders of Smo: SMANT compounds

based on the knowledge that different Smo antagonists affected Smo localization in different ways. As examples, SANT-1, SANT-2, and GDC-0449 inhibited Smo accumulation at the primary cilium, thus blocking Hh pathway activation (the accumulation of Smo within the primary cilium is a mandatory condition for the activation of Hh signaling). Differently, although cyclopamine and several of its derivatives were able to inhibit Hh pathway activation, they showed a pseudoagonist profile, promoting Smo accumulation at the primary cilium. A high content screen was then set up to assay a database of about 5,600 known compounds and to identify small molecules able to block Smo accumulation in the presence of Shh [99]. Upon elimination of compounds that blocked Hh pathway by non-specific mechanisms [as examples, the Hedgehog Pathway Inhibitor 4 (HPI-4, **193**, Fig. 24) [100] led to loss or truncation of the primary cilium, while vinblastine disrupted microtubule assemblies], 26 hits were identified. Among them, already known Smo antagonists were found, such as the quinazolinone derivative **34** [42], the antifungal drugs itraconazole and ketoconazole, as well as inhibitors of cholesterol biosynthesis (such as AY9944, **141**, that supports the hypothesis of a link between cholesterol and Hh pathway, Fig. 17). Two additional compounds [DY131 (**142**) and GSK4716 (**143**), already known as antagonists of the estrogen-related receptors] were also identified. They blocked ciliary accumulation of wild-type Smo consequent to both Hh ligand exposure and wild-type Smo overexpression, but failed in the case of a constitutively active W539L mutated form of Smo. Moreover, DY131 directly interacted with Smo, being a competitor of BODIPY-cyclopamine similar to cyclopamine, GDC-0449, and SANT-1. Differently, piperidine derivatives **144** and **145** (referred to as SMANT compounds) inhibited both wild-type and mutated Smo, although they were weak competitors of BODIPY-cyclopamine. Experimental evidence further suggested that SMANT acted at a binding site on Smo different from that targeted by other Smo inhibitors, such as cyclopamine, GDC-0449, and SANT-1 (the last compound was described as an allosteric Smo

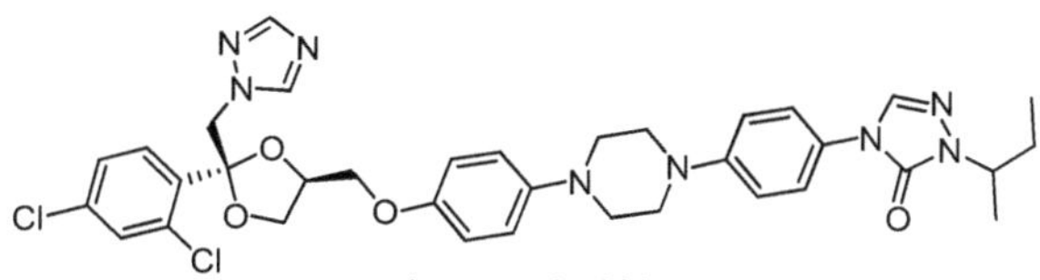

itraconazole, **146**

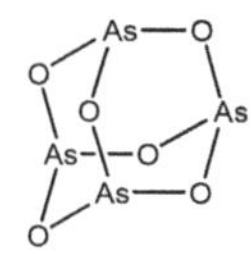

arsenic trioxide, **147**

Fig. 18 Itraconazole, a known antifungal agent, and arsenic trioxide, a drug for APL, have activity toward Smo and Gli proteins, respectively

inhibitor). IC_{50} values of DY131 and SMANT were measured in various cell-based assays (inhibition of Shh-induced Smo ciliary accumulation and inhibition of a Gli-Luc reporter also in Smo overexpressing cells) and were in the nanomolar range [99].

3.4 Itraconazole and Arsenic Trioxide: A Combined Activity Toward Smo and Gli Proteins

Itraconazole (**146**, Fig. 18) is an approved antifungal agent, while arsenic trioxide (**147**) is a drug approved for the treatment of acute promyelocytic leukemia (APL). Both of them are well characterized in terms of pharmacokinetics and toxicity profiles. Itraconazole is able to block Hh pathway activity by direct interaction with a binding site on Smo different from that occupied by cyclopamine and its mimics, because itraconazole does not compete with BODIPY-cyclopamine and the Smo agonist SAG [101]. In agreement, itraconazole is able to block all of the Smo mutants resistant to GDC-0449 and LDE225. Itraconazole-dependent Hh pathway inhibition is also independent from its antifungal target (lanosterol-14α demethylase). On the other hand, arsenic trioxide interacts directly with the causative agent of APL (the promyelogenous leukemia-retinoic acid receptor fusion protein) thus leading to its degradation. It is also a potent inhibitor of the Hh signaling by preventing Gli2 ciliary accumulation and inducing its degradation [102]. Arsenic trioxide directly binds to Gli1 and consequently inhibits the growth of cancer lines with upregulated Gli1 expression (such as Ewing sarcoma) [103]. Recent studies showed that arsenic trioxide and itraconazole, alone and in combination, are effective in inhibiting Hh pathway activity by targeting wild-type Smo, as well as Smo mutants generated by treatment with currently available Smo inhibitors [104]. Moreover, they are also effective toward Hh activity consequent to Gli2 overexpression. In vitro and in vivo assays showed that these compounds inhibited growth of cultured cells and tumors bearing both wild-type and drug-resistant mutated Smo. The combined activity toward Smo and Gli proteins could reduce the occurrence of resistant cell lines, in addition to allow for lower doses and adverse effects.

3.5 Compounds Acting at the Oxysterol Binding Site: Azacholesterol and Budesonide

A recent study on oxysterol derivatives led to the synthesis of 22-azacholesterol (**207**, Fig. 26) that was able to bind the oxysterol binding site of Smo located within the cysteine-rich domain of the protein (also referred to as Site B, see Sect. 6.4). As a consequence, **207** blocked the Hh signaling without interfering with other Smo inhibitors [105]. In a similar way, budesonide (**208**; Fig. 26), an anti-inflammatory glucocorticoid, bound to a conserved pocket of the cysteine-rich domain of a recombinant human Smo [106], and showed similar inhibitory properties toward both the wild-type and D473H (GDC-0449-resistant) Smo [107].

4 Compounds Affecting the Hh Signaling Pathway Without Direct Evidence of Smo Binding

4.1 Bis-amide Compounds (AstraZeneca)

Similarly to AMGEN, a cell-based Gli-Luc assay performed by AstraZeneca resulted in the identification of the bis-amide **148** (Table 14) that showed a 450 nM activity in the firefly reported assay, but low aqueous solubility and strong plasma protein binding (about 1 μM and 99%, respectively) [108]. Moreover, probably due to the presence of a bis-amide system, this compound also showed a 25 nM inhibition of the p38α mitogen-activated protein kinase (MAPK). Following the same design approach applied by AMGEN, the benzamide systems on the left molecular portion were replaced by five- and six-membered nitrogen-containing heterocycles. This structural change was also predicted to decrease compound affinity toward p38 protein, since the amide group of ligands is known to interact with a couple of acidic residues on p38. As a result, although most of the heterocycle derivatives gained nanomolar activity in both firefly reporter assay and the Hh-dependent differentiation assay (**149** as an example), several compounds were also characterized by better solubility (up to 350 μM) and significantly reduced inhibitory activity toward p38α (>100 μM). Among imidazole, pyrazole, thiazole, pyridine, pyrimidine, and purine derivatives, the fist ones emerged for their improved activity toward the Hh pathway, reduced off-target effect toward p38α, and increased solubility. As an example, the dimethyl imidazole derivative **150** showed nanomolar activity in both firefly and functional assay (6 and 11 nM, respectively), the best solubility (350 μM), and the lowest IC_{50} toward p38 (>100 μM).

Table 14 Bis-amide compounds identified by AstraZeneca

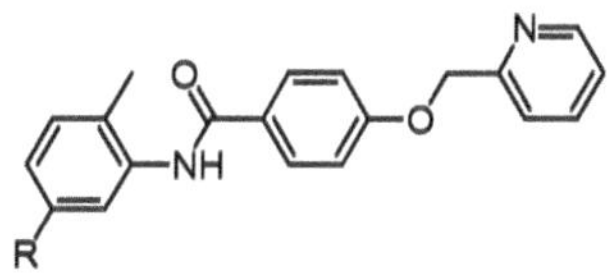

Comp	R	Firefly IC_{50} (nM)	Shh EC_{50} (nM)	p38α inhibition (IC_{50}, nM)	Solubility (μM)
148	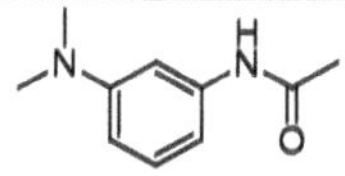	450		25	<1
149	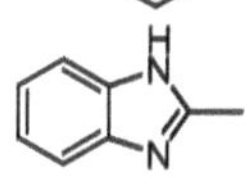	5	30	30	<1
150	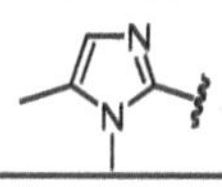	6	11	100	350

4.2 Pyrimidinyl-Amino-Benzamides and Pyrrolo-triazines (Jiangsu Simcere Pharmaceutical)

Merging structural portions of compounds already known to affect the Hh pathway, the pyrimidinylamino benzamide derivative **151** (Table 15) was designed and found to have a nanomolar IC_{50} (1.3 nM) [109] in a Gli-Luc reporter cell line (NIH3T3) previously described [73]. The overall structure of this compound is very similar to that of BMS-833923 (XL-139, **103**) [70], that is however a quinazoline derivative instead of a pyrimidinyl compound. Moreover, the trifluoromethoxy group at para position is also present in both LDE-225 (**92**) [61] and ALLO2 (142) [98]. Activity of **151** was better than that of both GDC-0449 (7.2 nM) and LDE-225 (5.5 nM). SAR studies suggested that deleting the trifluoromethoxy group or moving it to the meta position did not affect activity (1.3 and 1.2 nM, respectively). On the other hand, a two-fold improvement was found for the corresponding pyridine analogue **152** (0.5 nM). Moreover, introduction of cyclic and acyclic basic side chains at the position 5 of the xylenyl moiety improved activity [110]. As examples, morpholino (**153**), thiomorpholino, pyrrolidino, and *N*-alkylpiperazino derivatives were characterized by subnanomolar activity (0.5, 0.3, 0.4, 0.8 nM, respectively), as well as a series of acyclic tertiary amine derivatives (IC_{50} ranging from 0.2 to 0.9 nM). In vivo pharmacokinetic properties of these basic compounds were also very promising. As an example, oral bioavailability of the morpholino derivative was 46% in comparison with a 36% found for **151**. From this series of compounds, combination of activity (IC_{50} = 1.3 nM) and pharmacokinetic parameters of the pyrazole derivative **154** suggested it as a possible lead candidate to be tested in allograft models of medulloblastoma. In an attempt to further enhance activity toward Hh signaling pathway, the pyrimidine ring was transformed in a pyrrolo-triazine skeleton [111],

Table 15 Pyrimidinyl-amino-benzamides and pyrrolo-triazines identified by Jiangsu Simcere Pharmaceutical

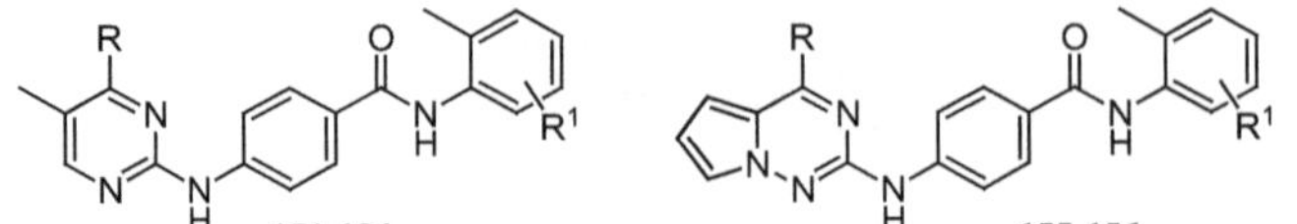

Comp	R	R^1	Gli-Luc reporter assay (IC_{50}, nM)
151	4-OCF_3-Ph	2-Me	1.3
152	4-pyridyl	2-Me	0.5
153	4-OCF_3-Ph	3-Morpholin-4-yl-methyl	0.5
154	1-Pyrazolyl	3-Morpholin-4-yl-methyl	1.3
155	4-OCF_3-Ph	2-Me	1.6
156	4-OCF_3-Ph	3-Morpholin-4-yl-methyl	0.8

reminiscent of the quinazoline nucleus of BMS-833923 (XL-139). The parent compound of this new series of pyrrolo[2,1-*f*][1,2,4]triazines **155** showed an activity very similar to that found for **151** (1.6 vs 1.3 nM, respectively). In agreement with previous SAR considerations, insertion of a basic side chain led to a slight improvement in activity to the subnanomolar range. The best compound **156** ($IC_{50} = 0.8$ nM) had satisfactory pharmacokinetic profile in comparison with GDC-0449 and was submitted to in vivo evaluation for solid tumor models.

5 Modulators of Hh Signaling Pathway Upstream and Downstream of Smo

5.1 Macrocyclic Inhibitors of ShhN (Broad Institute of Harvard and MIT)

A library of about 10,000 compounds from natural sources or obtained by DOS was used to perform a small-molecule microarray-based screen. These compounds were attached covalently on a glass surface and assayed for their ability to bind ShhN. As a result, a 13-membered macrocycle (**157**, Fig. 19) was reported as the first molecule able to bind the ShhN protein [112]. It showed moderate Hh pathway inhibition in a Gli-Luc assay, corresponding to about 30% of the inhibitory activity of cyclopamine. In the attempt to optimize its potency, changes on the decorating groups at the macrocyclic core led to compounds with a significant inhibition of Gli expression. As an example, **158** showed a 15 μM IC_{50} [113]. Moreover, reducing the size of the macrocycle to a 12-membered ring allowed the identification of the most active compound among all the macrocyclic analogues (**159**). Such a compound was named robotnikinin, from the name of Robotnik, the main antagonist of the Sonic the Hedgehog, a videogame character. Robotnikinin showed ability to

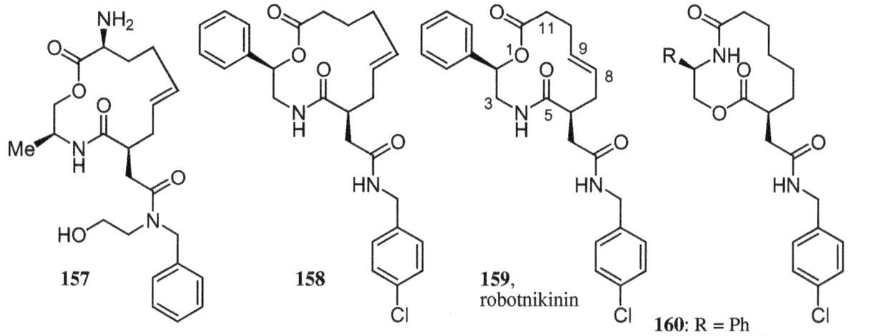

157 158 159, robotnikinin

160: R = Ph
161, BRD0607: R = Bn
162, BRD6851: R = *p*-Cl-Ph

Fig. 19 Macrocyclic compounds that act on ShhN

bind ShhN at a micromolar concentration. It did not compete with BODIPY-cyclopamine binding to Smo and did not affect Hh pathway in Ptch1-negative cells, thus suggesting a molecular target different from Smo and upstream of Ptch1. A new mechanism of action was proposed for robotnikinin, based on direct targeting of extracellular ShhN. When assayed for its ability to block differentiation of Hh-stimulated C3H10T1/2 cells into osteoblasts (measured as reduction in AP activity), robotnikinin showed an IC_{50} higher than 25 μM and a 47% maximal inhibition in comparison with cyclopamine. Changing the substituents on the macrocyclic core led to several compounds with improved potency [114]. As examples, 3-(*R*)- and 3-(*S*)-methyl derivatives of robotnikinin showed IC_{50} values of about 15 μM (69–77% maximal inhibition), as well as the corresponding 3-4-fused pyrrolidine analogue. Better results ($IC_{50} = 5$ μM and 86% maximal inhibition) were obtained for **160** bearing macrocycle oxygen and nitrogen in inverted positions in comparison with robotnikinin. Compounds of similar profile were found by its methylation or benzylation at position 11, and by transformation of the C8–C9 *E*-olefin into the corresponding *Z*-isomer. Further exploration of the 2-position of **160** showed that the phenyl ring (with *R* stereochemistry) could be profitably replaced by a *p*-F phenyl, benzyl (**161**, referred to as BRD0607), and cyclohexyl chain to maintain a single-digit micromolar IC_{50} and a maximal inhibition ranging from 75% to 90%. Unexpectedly, the *p*-Cl analogue (**162**, referred to as BRD6851) was about ten-fold more active ($IC_{50} = 0.5$ μM) than the parent compound **160** and showed a maximal inhibition comparable to that of cyclopamine. Very interestingly, by evaluation of Shh-induced Gli1 expression in SAG-stimulated C3H10T1/2 cells, evaluation of Gli1 transcription in Ptch-negative fibroblasts, and evaluation of competitive binding toward BODIPY-cyclopamine, both BRD0607 and BRD6851 were demonstrated to target directly Smo, differently from robotnikinin that is a direct binder of Shh.

In another study, the same authors applied a classical cell-based HTS on Shh-light2 cells to filter a database comprised of about 22,000 compounds previously obtained by DOS. A preliminary filter (inhibition >65%) was applied to prune compounds with low activity, and the resulting 390 hits were further profiled for their toxicity and ability to inhibit differentiation of Hh-stimulated C3H10T1/2

Table 16 Tetrahydrobenzoxazocinone derivatives with unknown molecular target, discovered at the Broad Institute of Harvard and MIT

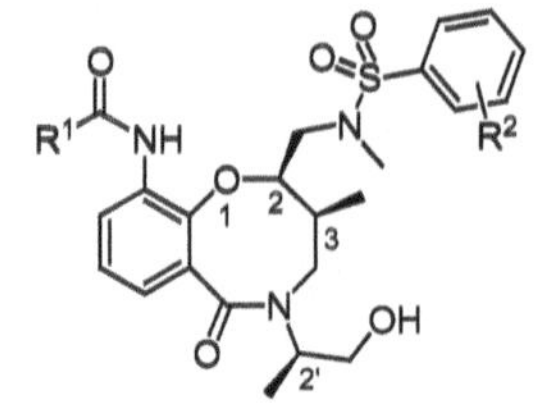

Comp	R^1	R^2	Solubility (μM)	Inhibition of Gli-induced transcription (EC50, nM)
163, *BRD50837*		*p*-Cl	64.3	90
164	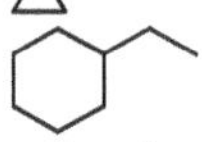	*p*-Cl	1.1	80
165	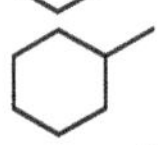	*p*-Cl	5.5	30
166	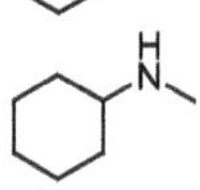	*p*-Cl	3.9	80
167, *BRD9526*		*o,p*-diCl	57.4	60

cells into osteoblasts. Tetrahydrobenzoxazocinone derivatives were identified whose activity strongly depended on their stereochemistry [115]. In particular, the 2-(*R*),3-(*S*),2′-(*R*) (**163**, BRD50837, Table 16) and the corresponding 2′-(*S*) stereoisomer showed a 90 and 440 nM IC_{50} in the C3H10T1/2 assay. A good solubility (about 64 μM) was also found BRD50837, that was thus chosen for further derivatization. Replacement of the cyclopropyl substituent with bulkier cycloalkyl groups and ureidic moieties was tolerated (**164–166**), while the solubility was significantly reduced to single-digit micromolar concentrations (from 1.1 to 5.5 μM). On the other hand, the substitution pattern of the chloride substituent on the terminal phenyl ring was a crucial key for activity. Removing the *p*-Cl or moving it to different positions led to compounds with activity one order of magnitude lower, at least. The *o,p*-dichloro analogue **167** (referred to as BRD9526) restored a 60 nM activity, further suggesting the importance of the chloride at para position. A deeper investigation of the mechanism of action of BRD50837 at the level of Hh signaling pathway showed peculiar results. It blocked SAG-induced C3H10T1/2 cell differentiation, similarly to cyclopamine. Moreover, Gli1 expression in SuFu-negative fibroblasts was partially inhibited, as also found for cyclopamine. Differently, Hh pathway of Ptch-negative cells was not inhibited by BRD50837, while cyclopamine did. These results did not provide striking evidence to fully understand the mechanism of action of such compounds. Authors suggested Ptch or its upstream Hh components or, in alternative, SuFu and its downstream components as a possible target for BRD50837 and BRD9526. Smo

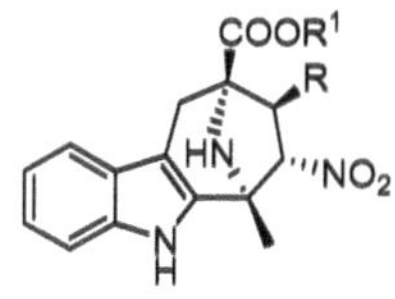

Fig. 20 Tropane derivatives able to inhibit AP enzymatic activity

was removed from the list of possible targets because BRD50837 and BRD9526 did not compete for BODIPY-cyclopamine binding on Smo.

A series of alkyl and aryl esters bearing a tropane nucleus (Fig. 20) was synthesized at the Max-Planck-Institute by [3+2] cycloaddition reaction and found to inhibit enzymatic activity of AP in C3H10T1/2 cells at micromolar concentrations [116].

5.2 Inhibitors of Glioma-Associated Oncogene Homologue 1 and 2 (Gli1 and Gli2)-mediated Transcription

5.2.1 Natural Compounds

A cell-based screening assay performed on tropical plant (*Zizyphus cambodiana*, Rhamnaceae) methanol extracts identified three structurally related pentacyclic triterpenes (namely, colubrinic acid **168**, betulinic acid **169**, and alphitolic acid **170**, Fig. 21) that showed a two-digit micromolar ability to inhibit Gli1-mediated transcriptional activity [117]. They also had similar cytotoxicity toward human pancreatic (PANC1) and prostate (DU145) cancer cells, associated with a decreased expression of the anti-apoptotic protein Bcl2. A similar biological profile was found for steroids **171** (Fig. 21) extracted from both *Adenium obesum* (Apocynaceae) [118] and *Vallaris glabra* (Apocynaceae) [119].

Gli1-reporter activity in Shh-light2 cells and growth of human PC3 and LNCaP prostate cancer cells were also inhibited by a methanol extract of *Sutherlandia frutescens* (Fabaceae, known as cancer bush in South Africa) [120].

Several flavonoids including genistein, curcumin, epigallocatechin 3-gallate, and resveratrol were described as Hh inhibitors able to decrease Shh-stimulated Gli reporter activity in Shh-light2 cells at micromolar concentrations, and to reduce prostate cancer in vivo in mice. Their mechanism of action was not further investigated [121, 122].

Additional natural compounds (Fig. 21), such as zerumbone (**172**), physalin B (**173**), physalin F (the 5–6 epoxide derivative of physalin B, **174**), staurosporinone (staurosporine aglycone, **175**), arcyriaflavin A (**176**), arcyriaflavin C (a dihydroxy derivative of arcyriaflavin A, **177**), inhibited both Gli1- and Gli2-mediated transcription, as well as repressed the transcription of other Shh signaling genes (*Ptch1* and *Bcl2*) in human keratinocyte cells. Although their mechanism of action is not

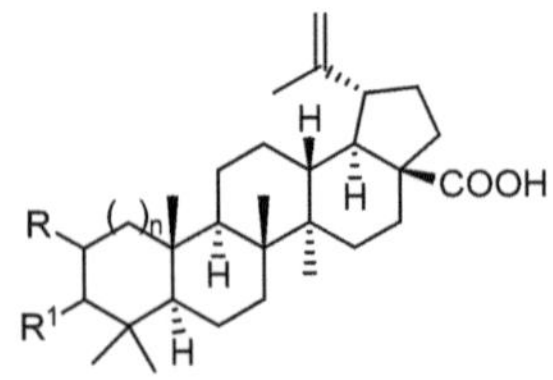

Colubrinic acid, **168**: n = 0, R = ▬CHO, R^1 = ┅OH
Betulinic acid, **169**: n = 1, R = H, R^1 = ▬OH
Alphitolic acid, **170**: n = 1, R = ┅OH, R^1 = ▬OH

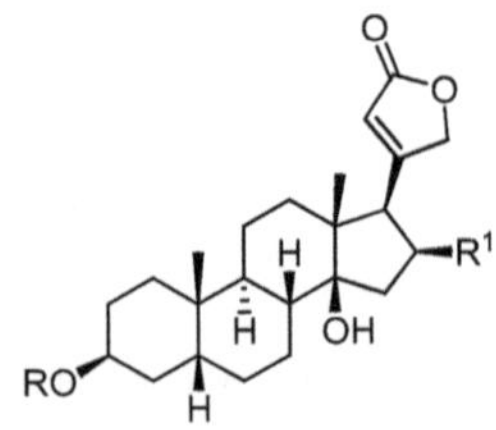

Alkaloids from *Adenium obesum* and *Vallaris glabra*
171

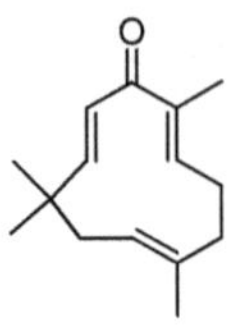

zerumbone, **172**

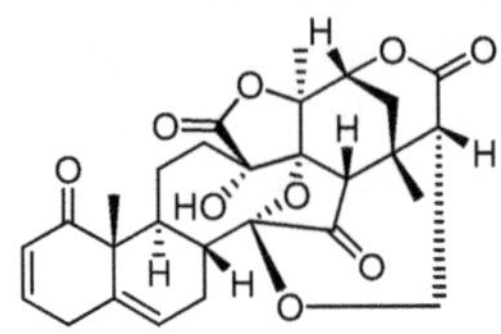

physalin B, **173**

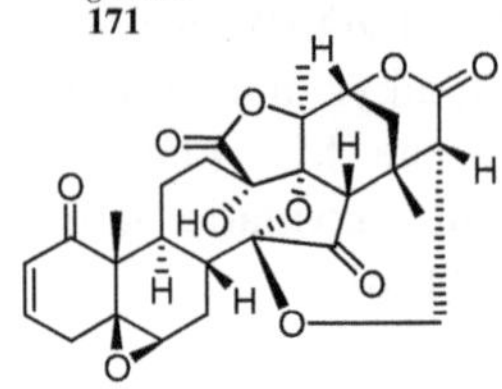

physalin F, **174**

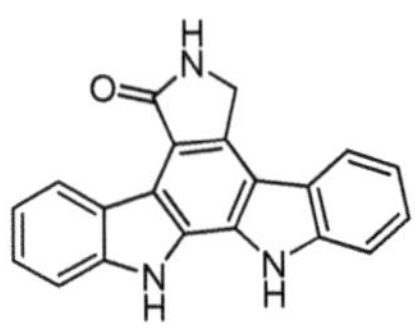

staurosporinone
staurosporine aglycone
175

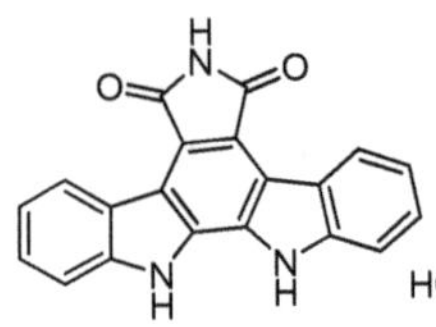

arcyriaflavin A, **176**

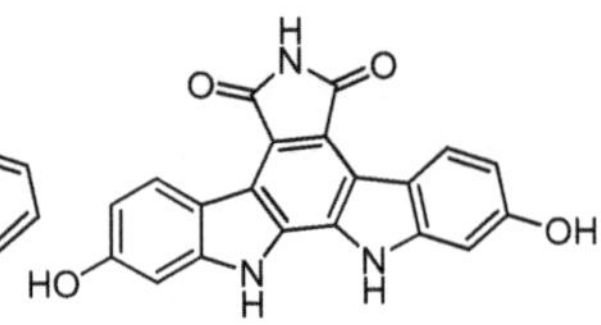

arcyriaflavin C, **177**

Fig. 21 Natural compounds from plants as inhibitors of Gli1 activity

well understood yet, it has been suggested that arcyriaflavin C and physalin F indirectly antagonized Gli function through protein kinase C (PKC)/MAPK pathway blockade [123]. Additional bis-indole derivatives showed micromolar inhibitory activity of Gli-mediated transcription and cytotoxicity toward both pancreatic and prostate tumor cells (namely PANC1 and DU145) without affecting normal C3H10T1/2 cells [124].

5.2.2 Benzophenone Derivatives Acting at Gli1 (St. Jude Children's Research Hospital)

Although most of the small molecule agonists and antagonists of Hh signaling pathway have Smo as the common molecular target, activation of the Hh-Gli signaling depends in some cases from regulators and transcription factors that are downstream of Smo (such as SuFu, Ren, Gli1-2). As a consequence, Smo inhibitors and compounds affecting those components that are upstream of Smo are ineffective in blocking the signaling pathway. On the contrary, compounds able to affect the final steps of the signal transduction will be able to inhibit the pathway, irrespective of the component that caused the activation of the Hh signaling, even if it corresponds to Hh ligands (Sonic, Desert, and Indian) or receptors (Ptch1 and Ptch2). In this context, C3H10T1/2 cells exogenously transfected with vectors for Gli1 and a Gli-Luc reporter were used to evaluate selective inhibition of Gli1-mediated transcription by small molecules. Such assay allowed the identification of only those compounds that are inhibitors of the Gli1-mediated transcription, which constitutes the final step in the Hh pathway, but not of upstream components of the same pathway [125]. In fact, cell line used in this assay is resistant to inhibition by cyclopamine and Hh-Antag, while sensitive to the Gli Antagonist 61 (GANT-61), a non-selective inhibitor Gli-mediated transcription. As a result, the pyrazoline derivative FN1-8 (**178**, Table 17) was identified as a small molecule able to target Gli proteins in human cancer cells [126]. Following previous SAR analysis, the tiramine moiety of this compound was kept as an essential pharmacophoric feature, while the triphenyl-pyrazoline scaffold was replaced with different aromatic groups [125]. While benzamide and benzylamide derivatives were all inactive, compounds with bulkier aromatic groups connected with a methylene bridge to the amide linker showed significant inhibition of the Gli1-mediated transcription. In particular, **179** and **180** strongly inhibited Gli1 activity, reducing it to 10 and 20%, respectively. Unfortunately, the latter compound was discarded because of its cytotoxicity. Further studies on the phenolic portion suggested that slight structural changes are tolerated. In fact, only the catechol analogue **181** showed better activity (more than 90% inhibition). Finally, keeping fixed both the methylene carbon atom in its (*S*)-configuration and the phenol terminal group, variations of the amide spacer led to less active compound, with the exception of urea derivatives (such as **182**) and N-substituted amides (Gli inhibition >90%). Urea compounds were however discarded for their toxicity, while substituted amides **183** (also referred to as NMDA298-1) and **184** were further profiled. Apart from the catechol derivative that underwent a rapid metabolic degradation, the remaining compounds were more stable and showed 3-4-fold selectivity in inhibiting Gli1-mediated transcription in comparison with Gli2 in C3H10T1/2 cells. They also inhibited endogenous Gli activity in a rhabdomyosarcoma cell line, thus suggesting their ability to block cancer growth dependent from Gli1 activity. This hypothesis was convincingly demonstrated also for several breast and prostate cancers, as well as for medulloblastoma and lymphoma cells.

To improve metabolic stability of the phenol derivatives, phenol group was replaced by nitrogenated heterocycles [127]. Indole derivatives showed best activity profile. In particular, **185** and **186**, bearing an indole group, inhibited Gli1-Luc reporter with a micromolar activity (IC_{50} 2.6 and 1.6 μM, respectively) and a 10-fold selectivity in comparison with inhibition of Gli2-mediated transcription.

5.2.3 GANT Compounds Acting at Gli1-2 (Karolinska Institutet)

A cell-based screening using HEK293 cells transiently expressing Gli1 and a luciferase reporter was set up at Karolinska Institutet to identify inhibitors of Gli1-mediated transcription. As a result, a hexahydropyrimidine (GANT-61, NSC136476, **187**) and a thiophene (GANT-58, NSC75503, **188**) derivative were

Table 17 Indole and Benzophenone derivatives discovered at St. Jude Children's Research Hospital, leading to the identification of NMDA298-1

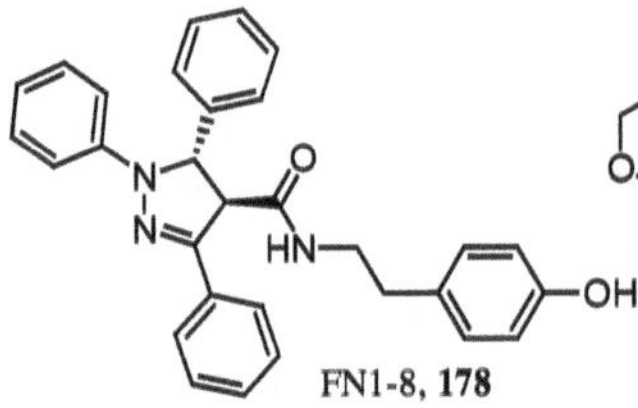

FN1-8, **178**

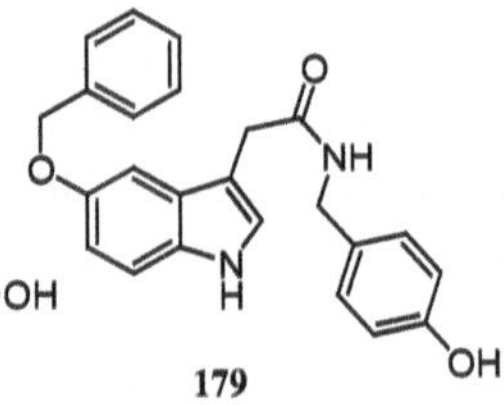

179

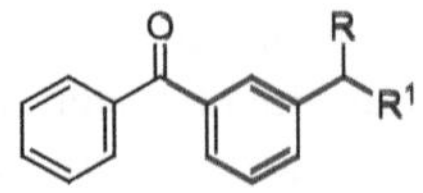

180-186

Comp	R	R^1	Gli1/Gli-Luc inhibition (%)
179			90
180	(*S*)-Me	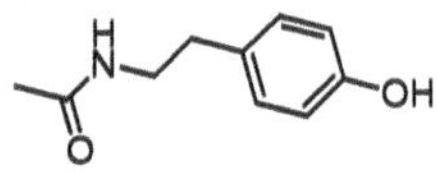	80
181	(*S*)-Me	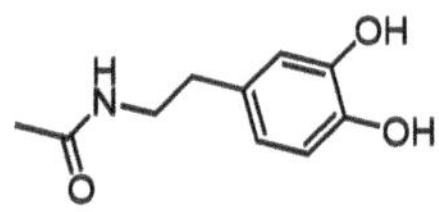	>90
182	(*S*)-Me	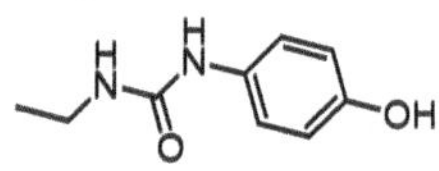	>90
183, *NMDA298-1*	(*S*)-Me	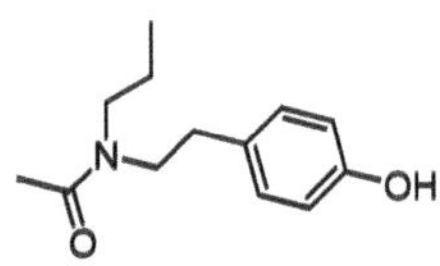	90
184	(*S*)-Me	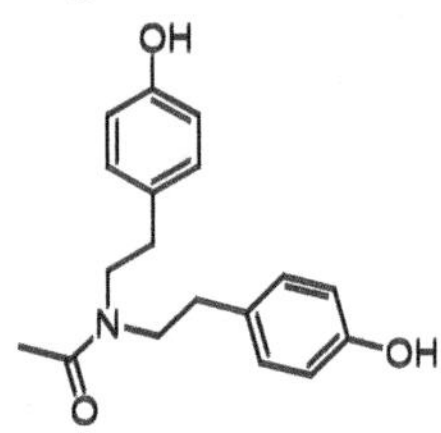	>90
185	Me	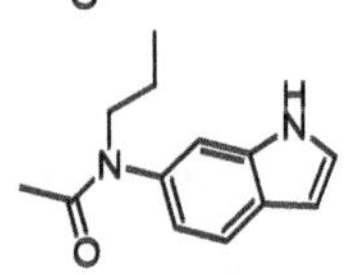	90
186	Me	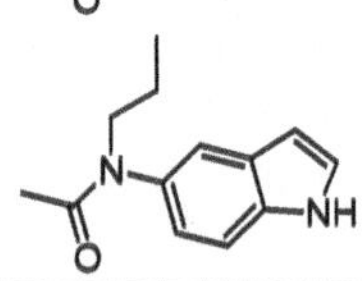	90

found (Fig. 22), both of them able to affect Gli1- and Gli2-mediated transcription with an IC_{50} of about 5 μM [128]. These compounds acted at the nucleus and inhibited Gli1 transcriptional activity. In particular, GANT-61 caused nuclear Gli1 accumulation and inhibited DNA binding of Gli1, either by preventing DNA

Fig. 22 GANT compounds act at Gli1-2

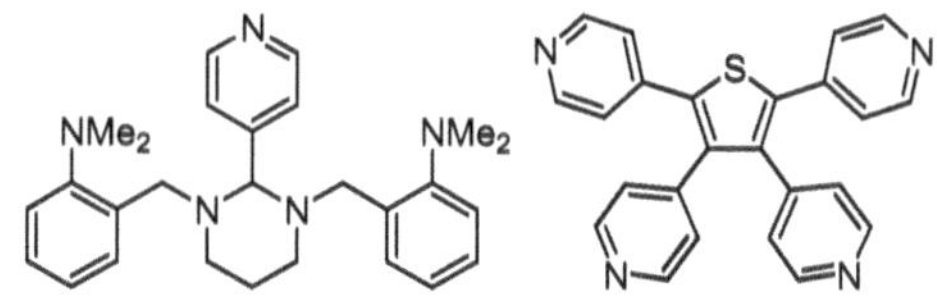

GANT61, NSC136476, **187** GANT58, NSC75503, **188**

Fig. 23 JK184 interferes with Gli2

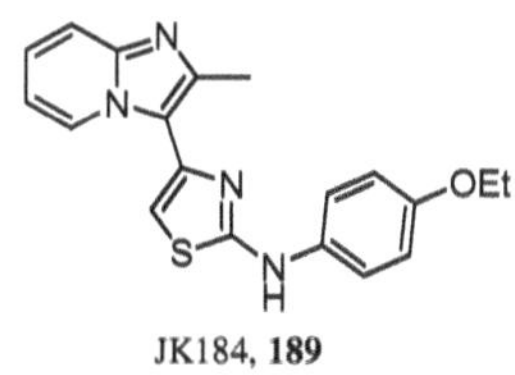

JK184, **189**

binding or by destabilizing DNA-Gli1 complex. GANT compounds also blocked in vitro Gli-dependent growth of a panel of human cancer cell lines, as well as an in vivo xenograft model of human prostate cancer. GANT-61 induced complete tumor regression, with a 50% recurrence after treatment discontinuation.

5.2.4 The Imidazo-Pyridine JK184 (Sankyo Company) Affects Gli2

ShhN-activated C3H10T1/2 cells treated with a library of about 20,000 compounds allowed the identification of the already known imidazo-pyridine derivative (JK184, **189** Fig. 23) [129] that inhibited Gli-dependent transcription with a 30 nM IC_{50} [130]. In in vitro experiments, a panel of cancer cell lines with abnormal activation of Hh signaling showed a decreased cell growth with GI_{50} values ranging from 3 to 20 nM. In vivo activity of JK184 was assayed in xenograft models of human pancreatic cancer, resulting in a 30–50% reduction of tumor growth. Poor pharmacokinetic parameters (a 1.3 h half-life and a 16% bioavailability) could be responsible for a superior in vitro profile, in comparison with in vivo activity. Affinity chromatography based on a JK184-coupled matrix led to the identification of alcohol dehydrogenase 7 (Adh7) as the target of JK184. These results suggested that the inhibition of Adh7 by JK184 could be in part responsible also for the effects on Hh signaling. Involvement of Adh7 in Hh signaling was also demonstrated by the fact that other Adh7 inhibitors, such as cimetidine and ranitidine, were also able to affect Gli-dependent transcription. In a subsequent study, ability of JK184 to inhibit Adh7 was considered as an ancillary activity in the context of Hh pathway inactivation [131]. In fact, experimental evidence showed that JK184 directly interacted with tubulin subunits and prevented their polymerization into filaments. As a consequence of microtubule depolymerization, microtubule-dependent conversion of full-length Gli2 proteins into transcriptional activators did not occur, thus perturbing Hh pathway activation. Studies on JK184 and its derivatives [132] allowed to rule out several SAR considerations. While various arylamines or arylamide at position 2 of the thiazole were tolerated, those bearing *p*-alkoxy

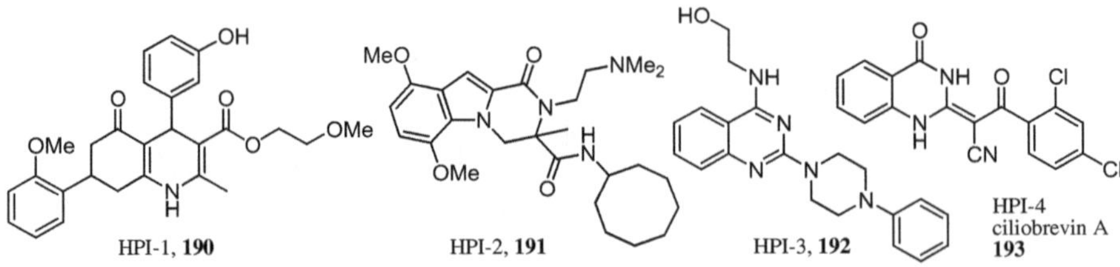

Fig. 24 HPI compounds affect Gli2 activity

substituents on the phenyl ring were preferred, although a significant preference did not exist for hydrophobic or hydrophilic groups. At position 4 of the thiazole ring, the imidazopyridine was a crucial structural feature.

5.2.5 HPI Compounds Act at Gli2 (Stanford University School of Medicine)

Within a research project aimed at identifying compounds that are able to block Hh pathway in Shh-light2 cells activated by SAG, four structurally different compounds were identified (thereafter referred to as Hh pathway inhibitors, HPI, **190–193**, Fig. 24) by screening of about 123,000 compounds [100]. Their biological profile was different from that of cyclopamine, in terms of both binding site (they did not compete with BODIPY-cyclopamine for its binding site on Smo) and functional activity in Shh-light2 cells (they were not functionally competitive with SAG). Moreover, each of the HPI showed a unique mechanism of action, different from that already found for synthetic (GANT compounds) [128] and naturally occurring (zerumbone, arcyriaflavin C, physalin F) [123] compounds acting downstream of Smo. Although such compounds partially inhibited Smo aggregation and affected Smo trafficking to the primary cilium, their action targeted Hh components that are downstream of Smo. HPI acted with a nonspecific mechanism that alters primary cilium structure. In particular, HPI-4 caused truncation or loss of the primary cilium [99], while HPI-1 and HPI-2 blocked Hh pathway activated by Gli overexpression or by loss of SuFu. Moreover, HPI-1 and HPI-4 were able to inhibit the proliferation of cerebellar granule neuron precursors expressing an oncogenic Smo mutant (SmoM2), as well as HPI-2 and HPI-3 blocked Hh pathway activation in SmoM2 fibroblast. Subsequent SAR studies on HPI-4 identified its 5-, 6-, and 7-monochloro derivatives that were able to inhibit both Hh signaling and primary cilium formation [133]. On the basis of their activity, such compounds were referred to as ciliobrevins A–D, respectively. Differently, monochloro derivatives bearing a chlorine substituent at position 3 or 4 of the benzoyl moiety only blocked Hh gene expression, without affecting cilium formation. In addition to reduce Hh target gene expression, ciliobrevin A (HPI-4) and its 7-chloro derivatives (ciliobrevin D) caused an increase of ciliary Gli2 level, similarly to that found upon loss of dynein 2 heavy chain. These results led to demonstrate that such compounds affect various dynein 1- and 2-dependent processes probably by direct targeting dynein proteins.

Table 18 Benzothiophene derivatives described by Curis, leading to the identification of Hh-Ag and SAG

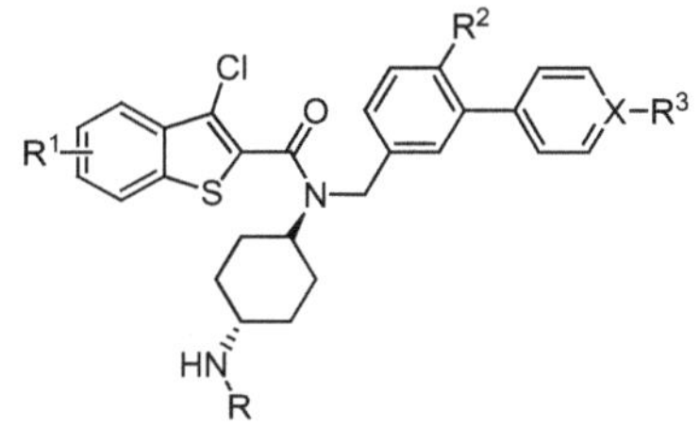

Comp	X	R	R[1]	R^2	R^3	Cell-based assay (EC_{50}, nM)
194, *Hh-Ag1.1*	CH	H	H	OMe	CN	5,000
cis-**194**	CH	H	H	OMe	CN	10,000
195, *Hh-Ag1.2*	CH	Me	H	OMe	CN	40
196	CH	Me	4,7-diF	OMe	CN	2
122, *Hh-Ag1.3, SAG*	N	Me	H	H		30
197, *Hh-Ag1.5*	N	Me	4,7-diF	H		1
198	N	Me	H	OMe		2
199	N	Me	4,7-diF	OMe		0.3

6 Compounds that Act as Hh Signaling Agonists

6.1 Benzothiophene Derivatives (Curis) Leading to Identification of Hh-Ag Compounds and SAG

The first example where Hh pathway was activated by small molecules independently of Hh endogenous ligands was represented by Hh-Ag compounds [47, 134] (also referred to as leiosamines) [135]. A cell-based assay setup with C3H10T1/2 cells bearing a luciferase reporter allowed to screen about 140,000 commercially available synthetic compounds and to identify agonists of the Hh pathway (Table 18). In particular, Hh-Ag1.1 (**194**, identified among the compounds purchased from Evotec-Oxford Asymmetry International) represented the parent compound with a half-maximal stimulation (EC_{50}) of about 5 μM and a 35% activation in comparison with the Hh protein used as control. In an attempt to find derivatives with improved potency, more than 300 derivatives of the parent compound were also assayed, then resulting in increasing activity up to three orders of magnitude. Hh-Ag1.5 (**197**) in fact showed an EC_{50} of about 1 nM. Both in vitro and in vivo experiments demonstrated that such compounds acted with a direct interaction on Smo. Moreover, their kinetic profile in comparison with known antagonists, such as cyclopamine, led to propose an allosteric binding with negative cooperativity, where Hh-Ag were bound to an active conformation of Smo while antagonists interacted with an inactive conformation of the same protein. Among the class of benzothiazole derivatives, Hh-Ag1.3 (SAG, **122**) showed the lowest toxicity toward several cell lines and was chosen (together with its difluoro analogue HhAg1.5) as one of the preferred reference compounds in many subsequent studies where it was referred to as SAG.

A larger series of SAG derivatives was also synthesized and tested in a Gli-Luc assay using C3H10T1/2 cells [136]. A SAR analysis clearly showed the pivotal role played by the *trans* geometry of the diaminocyclohexane moiety, the presence of an alkyl group at the terminal amine, and a chloro benzothiophene nucleus. A significant difference in activity was found between **194** (*trans* isomer, $IC_{50} = 5{,}000$ nM) and the corresponding *cis* isomer ($IC_{50} > 10{,}000$ nM), while the N-methyl analogue **195** was 125-fold more active (40 nM) than **194**. Decoration of the benzothiophene moiety of **195** by insertion of a fluoride at position 4, 5, or 7 led to comparable or better activity (15, 40, and 10 nM, respectively). Even better result was obtained with the 4,7-difluoro analogue **196** (2 nM). On the other hand, replacement of the *p*-cyanophenyl group with a 4-pyridyl substituent afforded **197–199** and **122** with activity up to the subnanomolar range. Further attempts to modify the structure of SAG did not result in better compounds [137].

6.2 *Purmorphamine (Scripps Research Institute and Genomics Institute of the Novartis Research)*

A convergent synthesis approach led to obtain about 50,000 compounds comprised of six-membered nitrogenated cyclic entries (pyrimidines, pyrazines) and the corresponding benzo-condensed derivatives (quinazolines, quinoxalines), in addition to purines and phthalazines [138]. Such compounds were assayed for their ability to induce differentiation of multipotent mesenchymal C3H10T1/2 progenitor cells into osteoblast, measured on the basis of AP enzymatic activity. A purinic derivative bearing a morpholino side chain (thereafter referred to as purmorphamine, **200**, Fig. 25) showed micromolar activity in activating both C3H10T1/2 and Shh-light2 cells, as well as low cytotoxicity. Further studies showed that purmorphamine competed with BODIPY-cyclopamine, thus providing evidence for a common binding site on Smo [95]. Two purmorphamine derivatives [KR96535 (**201**) and KR96547 (**202**)], identified with a differentiation assay on preosteoblast C2C12 mouse cells, upregulated *Ptch2* and *Gli2* genes involved in Hh signaling, but had opposite activity in inducing AP expression. Such compounds are presumed to act as de-differentiating agents that induce C2C12 cells to regress to multipotent mesenchymal progenitors [139] (as reversine **203**, a purmorphamine analogue, did) [140], although no experimental support was provided.

6.3 *Glucocorticoids as Smo Agonists*

A high-content screen using the human osteosarcoma U2OS cells led to identify four mono- and di-fluorinated glucocorticoids as Smo agonists within a library of

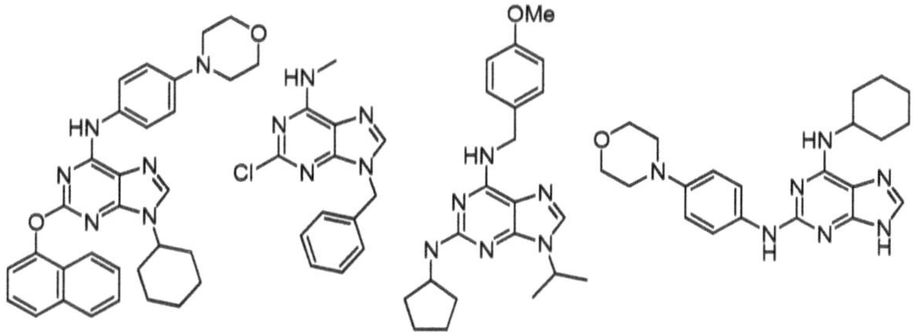

Purmorphamine, **200** KR96535, **201** KR96547, **202** Reversine, **203**

Fig. 25 Osteoinductive small molecules: purmorphamine and its derivatives

FDA-approved drug [141]. Halcinonide, fluticasone, clobetasol, and fluocinonide compete with BODIPY-cyclopamine in binding Smo, induce Smo internalization similarly to SAG and purmorphamine, activate Gli in Shh-light2 cells, and synergistically promote the proliferation of mouse primary neuronal precursor cells [142]. In a subsequent work [107], compounds able to modulate Smo accumulation at the primary cilium were filtered out from a library of known compounds (about 5,700 approved drugs and drug candidates). The most populated class was represented by natural and synthetic glucocorticoids. As an example among them, fluocinolone acetonide (the deacetylated analogue of fluocinonide already identified as Smo agonist [142]) promoted Smo accumulation without any significant Hh pathway activation. Interestingly, this compound sensitized cells to Shh-induced Hh pathway stimulation and targeted directly Smo at the cyclopamine binding site. In a similar way, most of the glucocorticoids identified shared the same binding site on Smo. On the contrary, budesonide (**208**, Fig. 26) and structurally similar glucocorticoids inhibited Smo ciliary accumulation and acted on Smo through a mechanism of action different from that found for many agonists and antagonists (that involves the cyclopamine binding site). The binding site of budesonide was subsequently identified in the cysteine-rich domain of Smo [106].

6.4 *Oxysterols are Agonists of the Hh Signaling Pathway by Allosteric Activation of Smo*

Oxysterols were demonstrated to activate Hh signaling by inducing accumulation of Smo in the primary cilium. Hh activation by oxysterols is stereoselective [143]. In fact, while the natural 20-(*S*)-hydroxy cholesterol (**204**, Fig. 26) was the most potent activator of Hh pathway (with a half-maximal activation reached at micromolar concentrations), both the 20-(*R*) epimer (**205**, one of the 256 diastereoisomers) and the enantiomer (a compound with all the seven stereocenters in the opposite stereochemistry) were inactive. The natural compound **204** showed a synergy with SAG that suggested a positive allosteric interaction on Smo. This means that **204** binds to a Smo site (referred to as Site B, see below) different from

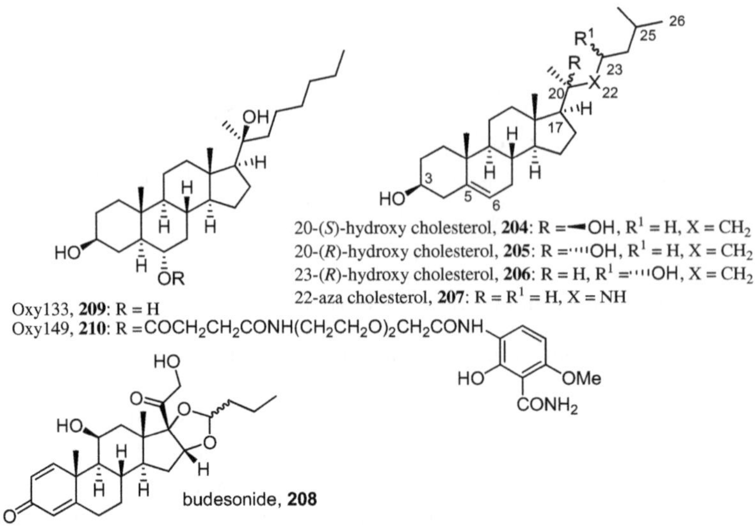

Fig. 26 Oxysterols that act as agonists of the Hh signaling pathway. Budesonide and 22-aza cholesterol are Smo inhibitors that act at the CRD of Smo

that accommodating both SAG and cyclopamine (referred to as Site A, see below). This is also in agreement with the non-competitive interaction between cyclopamine and **204**. However, the latter showed probe specificity because it had competitive interactions with SANT-1 and SANT-2, that are Smo antagonists able to compete with cyclopamine for binding to Smo.

In an attempt to find osteogenic oxysterol suitable for clinical development, more than 100 oxysterol analogues were synthesized and tested for their ability to induce expression of osteogenic markers in mouse C3H10T1/2 cells. Among the resulting compounds, Oxy133 (**209**, Fig. 26) was able to activate Hh pathway by direct interaction with Smo and its activity is antagonized by cyclopamine [144]. Oxy149 (**210**), a derivative of Oxy133 bearing a structural motif that facilitates bone targeting, selectively accumulates in bone and enhances osteogenic differentiation by activation of Hh signaling [145]. Structure–activity relationship studies focused on the side chain at position 17 led to identify 23-(*R*)-hydroxy cholesterol (**206**) as a micromolar activator of Hh signaling with a significant selectivity in comparison with the liver X receptor (a nuclear receptor also activated by hydroxy cholesterol derivatives) [146].

The site of Smo that accommodates various small molecules, inhibitors or activators of Smo, has been recently referred to as Site A [105]. Differently, the allosteric site on Smo occupied by oxysterols has been termed Site B. In the attempt to shed light on the mechanism by which oxysterols and other Smo ligands are able to interact with their receptor, 22-azacholesterol (**207**, Fig. 26) was synthesized on the basis of the hypothesis that such a compound could be a potential inhibitor of the agonist activity of **204** on Smo. A micromolar inhibition of Hh signaling induced by both Shh and **204** was found upon treatment with **207**. As expected, competition for the oxysterol binding site on Smo was critically dependent from C20 stereochemistry, with the *S*-configuration significantly preferred. Binding of BODIPY-cyclopamine to Smo was not affected by **207**. Accordingly, Site B was found to be the common target of **204** and **207**. Interestingly, Site B is located within the extracellular cysteine-rich domain of Smo and is completely separate from Site A (accommodated within the heptahelical bundle) that is targeted by other small molecule agonists and antagonists. Experimental evidence suggested that a dual activation of Smo is required for Hh signaling: Site A is involved in the interactions with Hh ligand (namely, Shh), while Site B is occupied by an oxysterol (namely, **204**). From a functional point of view, Hh signaling activation seems to be dependent from Shh/Ptch, while oxysterol binding to Site B could act as a modulating factor. On this basis, blocking Site B with **207** or similar compounds could represent an alternative approach to inhibit Smo activity.

During the study that identified **207**, SAR considerations were ruled out on oxysterol analogues as Smo activators. In fact, shortening the C22–C27 tail to an *n*-pentyl and *n*-butyl chain maintained an activity comparable to that of **204**, while shorter chains were not profitable for activity. Moreover, the C5–C6 double bond was not necessary for activity, as already demonstrated by Oxy compounds. On the contrary, the 3β-hydroxy group is of crucial importance.

6.5 GSA-10 is a Quinolinone Activator of Smo by a Peculiar Mechanism of Action (University of Siena, CNRS Gif-sur-Yvette, Université de Strasbourg)

A quinolone derivative, referred to as GSA-10 (**211**, Fig. 27) [147], is the sole example of a novel small molecule positive modulator of Smo discovered by an in silico ligand-based virtual screening approach that applied a pharmacophoric model for filtering virtual databases of commercially available compounds. GSA-10 was profiled to dissect its biological activity at the molecular level and to fully understand its involvement in the Hh signaling pathway. In detail, **211** bound to Smo and stimulated differentiation of C3H10T1/2 cells into osteoblasts. However, while SAG competed with BODIPY-cyclopamine, promoted Smo accumulation at the primary cilium, induced a 60-fold increase of *Gli1* gene transcription, and stimulated a dose-dependent increase of granular cell precursor (GCP) proliferation, **211** showed an opposite effect. In fact, it was unable to compete with BODIPY-cyclopamine and to promote Smo accumulation. In addition, **211** caused a 3-fold decrease of gene transcription and did not induce GCP proliferation. Moreover, treatment with forskolin, a negative regulator of Hh signaling pathway, resulted in the inhibition of SAG-induced C3H10T1/2 differentiation, while evoking a synergistic effect in differentiation induced by **211** [147]. These results suggested that the two compounds are Smo agonists with a different mechanism of action, working through different transduction mechanisms. Variation in AP activity induced by SAG or **211** in the presence or the absence of the other compound provided strong

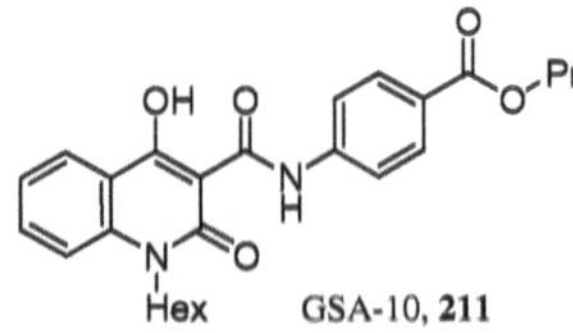

GSA-10, **211**

Fig. 27 GSA-10, a quinolinone agonist of Smo, discovered by virtual screening

evidence for a synergistic activity between these Smo agonists, suggesting that the two compounds could act independently on different Smo receptor sites. The existence of two different binding sites for agonists on Smo was further supported by the fact that MRT-83 (**128**, a potent Smo antagonist) [88, 89], LDE225, and SANT-1 were very efficient in blocking AP activity induced by both SAG and **211** (consistent with a direct interaction of both SAG and **211** with Smo for pathway activation), while GDC-0449, CUR61414, and cyclopamine were potent inhibitors of SAG-induced differentiation, but displayed a reduced ability to block differentiation induced by **211** [147].

Although a biological profile significantly different from that of the known Smo agonist SAG emerged, strong evidence for the direct interaction between **211** and Smo was provided. SAG and **211** were hypothesized to have different binding sites on Smo and to mediate osteoblast differentiation in a synergistic manner, but with a different mechanism of action. At the moment, a Smo-dependent non-canonical pathway with a mechanism of action different from that of glucocorticoids [142], which are able to bind to Smo, to activate *Gli* genes, and to promote mouse GCP cell proliferation, cannot be excluded.

7 Conclusions

In this review, we report an up-to-date survey of small molecule that is able to block or activate the Hh signaling pathway by interaction with Smo or upstream and downstream component of the pathway. There is clear evidence that modulation of the Hh signaling pathway could be very fruitful for anticancer therapy and regenerative medicine. In the last years, a significant number of small molecules that are able to block or to stimulate the Hh pathway by targeting various components have been discovered and studied in depth. Many of them directly interact with the Smo protein. One of the Smo-targeted compounds, GDC-0449, is now in the market (vismodegib) for treatment of BCC. Further efforts are, however, required to identify by high-throughput screening or by rational design next generation compounds to be associated with currently known Hh inhibitors or activators, in the attempt to circumvent the main issue of Smo-mutated cell line resistant to GDC-0449 and to other compound under clinical phase.

References

1. McMillan R, Matsui W (2012) Molecular pathways: the Hedgehog signaling pathway in cancer. Clin Cancer Res 18:4883–4888
2. Scales SJ, de Sauvage FJ (2009) Mechanisms of Hedgehog pathway activation in cancer and implications for therapy. Trends Pharmacol Sci 30:303–312
3. Amakye D, Jagani Z, Dorsch M (2013) Unraveling the therapeutic potential of the Hedgehog pathway in cancer. Nat Med 19:1410–1422
4. Yun JI, Kim HR, Park H, Kim SK, Lee J (2012) Small molecule inhibitors of the Hedgehog signaling pathway for the treatment of cancer. Arch Pharm Res 35:1317–1333
5. Peukert S, Miller-Moslin K (2010) Small-molecule inhibitors of the Hedgehog signaling pathway as cancer therapeutics. ChemMedChem 5:500–512
6. Tremblay MR, McGovern K, Read MA, Castro AC (2010) New developments in the discovery of small molecule Hedgehog pathway antagonists. Curr Opin Chem Biol 14:428–435
7. Mahindroo N, Punchihewa C, Fujii N (2009) Hedgehog-Gli signaling pathway inhibitors as anticancer agents. J Med Chem 52:3829–3845
8. Lin TL, Matsui W (2012) Hedgehog pathway as a drug target: smoothened inhibitors in development. OncoTargets Ther 5:47–58
9. Hadden MK (2014) Hedgehog pathway agonism: therapeutic potential and small-molecule development. ChemMedChem 9:27–37
10. Stanton BZ, Peng LF (2010) Small-molecule modulators of the Sonic Hedgehog signaling pathway. Mol BioSyst 6:44–54
11. Taipale J, Chen JK, Cooper MK, Wang B, Mann RK, Milenkovic L, Scott MP, Beachy PA (2000) Effects of oncogenic mutations in smoothened and patched can be reversed by cyclopamine. Nature 406:1005–1009
12. Hamon F, Renoux B, Chadeneau C, Muller J-M, Papot S (2010) Study of a cyclopamine glucuronide prodrug for the selective chemotherapy of glioblastoma. Eur J Med Chem 45:1678–1682
13. Renoux B, Legigan T, Bensalma S, Chadeneau C, Muller J-M, Papot S (2011) A new cyclopamine glucuronide prodrug with improved kinetics of drug release. Org Biomol Chem 9:8459–8464
14. Kumar SK, Roy I, Anchoori RK, Fazli S, Maitra A, Beachy PA, Khan SR (2008) Targeted inhibition of Hedgehog signaling by cyclopamine prodrugs for advanced prostate cancer. Bioorg Med Chem 16:2764–2768
15. Tremblay MR, Nevalainen M, Nair SJ, Porter JR, Castro AC, Behnke ML, Yu L-C, Hagel M, White K, Faia K, Grenier L, Campbell MJ, Cushing J, Woodward CN, Hoyt J, Foley MA, Read MA, Sydor JR, Tong JK, Palombella VJ, McGovern K, Adams J (2008) Semisynthetic cyclopamine analogues as potent and orally bioavailable Hedgehog pathway antagonists. J Med Chem 51:6646–6649
16. Feldmann G, Fendrich V, McGovern K, Bedja D, Bisht S, Alvarez H, Koorstra J-BM, Habbe N, Karikari C, Mullendore M, Gabrielson KL, Sharma R, Matsui W, Maitra A (2008) An orally bioavailable small-molecule inhibitor of Hedgehog signaling inhibits tumor initiation and metastasis in pancreatic cancer. Mol Cancer Ther 7:2725–2735
17. Tremblay MR, Lescarbeau A, Grogan MJ, Tan E, Lin G, Austad BC, Yu L-C, Behnke ML, Nair SJ, Hagel M, White K, Conley J, Manna JD, Alvarez-Diez TM, Hoyt J, Woodward CN, Sydor JR, Pink M, McDougall J, Campbell MJ, Cushing J, Ferguson J, Curtis MS, McGovern K, Read MA, Palombella VJ, Adams J, Castro AC (2009) Discovery of a potent and orally active Hedgehog pathway antagonist (IPI-926). J Med Chem 52:4400–4418
18. Lee MJ, Hatton BA, Villavicencio EH, Khanna PC, Friedman SD, Ditzler S, Pullar B, Robison K, White KF, Tunkey C, LeBlanc M, Randolph-Habecker J, Knoblaugh SE, Hansen S, Richards A, Wainwright BJ, McGovern K, Olson JM (2012) Hedgehog pathway

inhibitor saridegib (IPI-926) increases lifespan in a mouse medulloblastoma model. Proc Natl Acad Sci USA 109:7859–7864
19. Moschner J, Chentsova A, Eilert N, Rovardi I, Heretsch P, Giannis A (2013) Cyclopamine analogs bearing exocyclic methylenes are highly potent and acid-stable inhibitors of Hedgehog signaling. Beilstein J Org Chem 9:2328–2335
20. Winkler JD, Isaacs A, Holderbaum L, Tatard V, Dahmane N (2009) Design and synthesis of inhibitors of Hedgehog signaling based on the alkaloid cyclopamine. Org Lett 11:2824–2827
21. Guerlet G, Spangenberg T, Mann A, Faure H, Ruat M (2011) Synthesis and biological evaluation of desmethylveramiline, a micromolar Hedgehog inhibitor. Bioorg Med Chem Lett 21:3608–3612
22. Lipinski RJ, Dengler E, Kiehn M, Peterson RE, Bushman W (2007) Identification and characterization of several dietary alkaloids as weak inhibitors of Hedgehog signaling. Toxicol Sci 100:456–463
23. Bijlsma MF, Spek CA, Zivkovic D, van de Water S, Rezaee F, Peppelenbosch MP (2006) Repression of smoothened by patched-dependent (pro-)vitamin D3 secretion. PloS Biol 4: e232
24. Teichert AE, Elalieh H, Elias PM, Welsh J, Bikle DD (2011) Overexpression of hedgehog signaling is associated with epidermal tumor formation in vitamin D receptor-null mice. J Invest Dermatol 131:2289–2297
25. Banerjee U, Ghosh M, Hadden MK (2012) Evaluation of vitamin D3 A-ring analogues as Hedgehog pathway inhibitors. Bioorg Med Chem Lett 22:1330–1334
26. DeBerardinis AM, Banerjee U, Miller M, Lemieux S, Hadden MK (2012) Probing the structural requirements for vitamin D3 inhibitors of the Hedgehog signaling pathway. Bioorg Med Chem Lett 22:4859–4863
27. DeBerardinis AM, Banerjee U, Hadden MK (2013) Identification of vitamin D3-based Hedgehog pathway inhibitors that incorporate an aromatic A-ring isostere. ACS Med Chem Lett 4:590–595
28. DeBerardinis AM, Madden DJ, Banarjee U, Sail V, Raccuia DS, De Carlo D, Lemieux SM, Meares A, Hadden MK (2014) Structure–activity relationships for vitamin D3-based aromatic A-ring analogues as Hedgehog pathway inhibitors. J Med Chem 57(9):3724–3736
29. Baxter AD, Boyd EA, Guicherit OM, Price S, Rubin LL (2001) Mediators of the Hedgehog signaling pathways, compositions and uses related thereto. WO 26644 A2
30. Williams JA, Guicherit OM, Zaharian BI, Xu Y, Chai L, Wichterle H, Kon C, Gatchalian C, Nusse R, Porter JA, Rubin LL, Wang FY (2003) Identification of a small molecule inhibitor of the hedgehog signaling pathway: effects on basal cell carcinoma-like lesions. Proc Natl Acad Sci USA 100:4616–4621
31. Tang T, Tang JY, Li D, Reich M, Callahan CA, Fu L, Yauch RL, Wang F, Kotkow K, Chang KS, Shpall E, Wu A, Rubin LL, Marsters JC Jr, Epstein EH Jr, Caro I, de Sauvage FJ (2011) Targeting superficial or nodular basal cell carcinoma with topically formulated small molecule inhibitor of smoothened. Clin Cancer Res 17:3378–3387
32. Rubin L, Giucherit OM, Price S, Boyd EA (2003) Mediators of Hedgehog signaling pathways, compositions and uses related thereto. WO 011219 A2
33. Gunzner JL, Sutherlin D, Stanley MS, Bao L, Castanedo GM, LaLonde RL, Wang S, Reynolds ME, Savage SJ, Malesky K, Dina MS (2006) Pyridyl inhibitors of Hedgehog signalling. US 0063779 A1
34. Robarge KD, Brunton SA, Castanedo GM, Cui Y, Dina MS, Goldsmith R, Gould SE, Guichert O, Gunzner JL, Halladay J, Jia W, Khojasteh C, Koehler MFT, Kotkow K, La H, LaLonde RL, Lau K, Lee L, Marshall D, Marsters J Jr, Murray LJ, Qian C, Rubin LL, Salphati L, Stanley MS, Stibbard JHA, Sutherlin DP, Ubhayaker S, Wang S, Wong S, Xie M (2009) GDC-0449 – a potent inhibitor of the Hedgehog pathway. Bioorg Med Chem Lett 19:5576–5581
35. Castanedo GM, Wang S, Robarge KD, Balckwood E, Burdick D, Chang C, Dijkgraaf GJP, Gould S, Gunzner J, Guichert O, Halladay J, Khojasteh C, Lee L, Marsters JC Jr, Murray L,

Peterson D, Plise E, Salphati L, de Sauvage FJ, Wong S, Sutherlin DP (2010) Second generation 2-pyridyl biphenyl amide inhibitors of the hedgehog pathway. Bioorg Med Chem Lett 20:6748–6753

36. Rudin CM, Hann CL, Laterra J, Yauch RL, Callahan CA, Fu L, Holcomb T, Stinson J, Gould SE, Coleman B, LoRusso PM, Von Hoff DD, de Sauvage FJ, Low JA (2009) Treatment of medulloblastoma with hedgehog pathway inhibitor GDC-0449. N Engl J Med 361:1173–1178
37. Asklund T, Henriksson R, Axelsson J, Bergstrom A, Kasper M, Ogren M, Toftgard R, Riklund KA (2013) Early and persisting response to vismodegib in a patient with bone metastasizing medulloblastoma. Acta Oncol 52:862–866
38. Atwood SX, Chang AL, Oro AE (2012) Hedgehog pathway inhibition and the race against tumor evolution. J Cell Biol 199:193–197
39. Yauch RL, Dijkgraaf GJP, Alicke B, Januario T, Ahn CP, Holcomb T, Pujara K, Stinson J, Callahan CA, Tang T, Bazan JF, Kan Z, Seshagiri S, Hann CL, Gould SE, Low JA, Rudin CM, de Sauvage FJ (2009) Smoothened mutation confers resistance to a Hedgehog pathway inhibitor in medulloblastoma. Science 326:572–574
40. Dijkgraaf GJP, Alicke B, Weinmann L, Januario T, West K, Modrusan Z, Burdick D, Goldsmith R, Bobarge K, Sutherlin D, Scales SJ, Gould SE, Yauch RL, de Sauvage FJ (2011) Small molecule inhibition of GDC-0449 refractory smoothened mutants and downstream mechanisms of drug resistance. Cancer Res 71:435–444
41. Baxter AD, Boyd EA, Guicherit OM, Price S, Rubin LD (2001) Mediators of the Hedgehog signaling pathways, compositions and uses related thereto. WO 019800 A3
42. Brunton SA, Stibbard JHA, Rubin LL, Kruse LI, Guicherit OM, Boyd EA, Price S (2008) Potent inhibitors of the Hedgehog signaling pathway. J Med Chem 51:1108–1110
43. Thomas RJ, Pericot Mohr G, Minetto G, Bakker A, Ferruzzi P (2009) The hedgehog pathway antagonists and therapeutic thereof. WO 074300 A2
44. Tunici P, Mancuso M, Giordano C, Ferruzzi P, Mennillo F, Tanori M, Saran A, Thomas R, Pericot Mohr G, Bakker A (2011) Selective smoothened antagonists for anti-tumor therapy. In: Proceedings of the 102nd annual meeting of the American Association for cancer research; 2011 April 2–6; Orlando, FL. Philadelphia (PA): AACR; Cancer Res 2011, 71, Abstract number LB-38
45. Ferruzzi P, Mennillo F, De Rosa A, Giordano C, Rossi M, Benedetti G, Magrini R, Pericot Mohr G, Miragliotta V, Magnoni L, Mori E, Thomas R, Tunici P, Bakker A (2012) In vitro and in vivo characterization of a novel Hedgehog signaling antagonist in human glioblastoma cell lines. Int J Cancer 131:E33–E44
46. Dessole G, Branca D, Ferrigno F, Kinzel O, Muraglia E, Palumbi MC, Rowley M, Serafini S, Steinkuhler C, Jones P (2009) Discovery of *N*-[(1-aryl-1*H*-indazol-5-yl)methyl]amides derivatives as smoothened antagonists for inhibition of the hedgehog pathway. Bioorg Med Chem Lett 19:4191–4195
47. Frank-Kamenetsky M, Zhang XM, Bottega S, Guicherit O, Wichterle H, Dudek H, Bumcrot D, Wang FY, Jones S, Shulok J, Rubin LL, Porter JA (2002) Small-molecule modulators of Hedgehog signaling: identification and characterization of smoothened agonists and antagonists. J Biol 1:10
48. Chen JK, Taipale J, Cooper MK, Beachy PA (2002) Inhibition of Hedgehog signaling by direct binding of cyclopamine to smoothened. Genes Dev 16:2743–2748
49. Malancona S, Altamura S, Filocamo G, Kinzel O, Hernando JIM, Rowley M, Scarpelli R, Steinkuhler C, Jones P (2011) Identification of MK-5710 ((8a*S*)-8*a*-methyl-1,3-dioxo-2-[(1*S*,2*R*)-2-phenylcyclopropyl]-*N*-(1-phenyl-1*H*-pyrazol-5-yl)hexahydroimidazo[1,5-*a*]pyrazine-7(1*H*)-carboxamide), a potent smoothened antagonist for use in Hedgehog pathway dependent malignancies, Part 1. Bioorg Med Chem Lett 21:4422–4428
50. Kinzel O, Alfieri A, Altamura S, Brunetti M, Bufali S, Colaceci F, Ferrigno F, Filocamo G, Fonsi M, Gallinari P, Malancona S, Hernando JIM, Monteagudo E, Orsale MV, Palumbi MC, Pucci V, Rowley M, Sasso R, Scarpelli R, Steinkuhler C, Jones P (2011) Identification of

MK-5710 ((8*aS*)-8*a*-methyl-1,3-dioxo-2-[(1*S*,2*R*)-2-phenylcyclopropyl]-*N*-(1-phenyl-1*H*-pyrazol-5-yl)hexahydroimidazo[1,5-*a*]pyrazine-7(1*H*)-carboxamide), a potent smoothened antagonist for use in Hedgehog pathway dependent malignancies, Part 2. Bioorg Med Chem Lett 21:4429–4435
51. Ontoria JM, Llauger Bufi L, Torrisi C, Bresciani A, Giomini C, Rowley M, Serafini S, Bin H, Hao W, Steinkuhler C, Jones P (2011) Identification of a series of 4-[3-(quinolin-2-yl)-1,2,4-oxadiazol-5-yl]piperazinyl ureas as potent smoothened antagonist hedgehog pathway inhibitors. Bioorg Med Chem Lett 21:5274–5282
52. Muraglia E, Ontoria JM, Branca D, Dessole G, Bresciani A, Fonsi M, Giuliano C, Llauger Bufi L, Monteagudo E, Palumbi MC, Torrisi C, Rowley M, Steinkuhler C, Jones P (2011) N-(2-alkylaminoethyl)-4-(1,2,4-oxadiazol-5-yl)piperazine-1-carboxamides as highly potent smoothened antagonists. Bioorg Med Chem Lett 21:5283–5288
53. Balkovec JM, Thieringer R, Waddell ST (2008) Triazole derivatives which are Smo antagonists. WO 130552 A1
54. Austin RJ, Kaizerman JA, Lucas BS, McMinn DL, Powers JP (2009) Phthalazine compounds, compositions and methods of use. WO 002469 A1
55. Lucas BS, Aaron W, An S, Austin RJ, Brown M, Chan H, Chong A, Hungate R, Huang T, Jiang B, Johnson MG, Kaizerman JA, Lee G, McMinn DL, Orf J, Powers JP, Rong M, Toteva MM, Uyeda C, Wickramasinghe D, Xu G, Ye Q, Zhong W (2010) Design of 1-piperazinyl-4-arylphthalazines as potent smoothened antagonists. Bioorg Med Chem Lett 20:3618–3622
56. Kaizerman JA, Lucas BS, McMinn DL, Zamboni R (2009) Annelated pyridazines for the treatment of tumors driven by inappropriate Hedgehog signaling. WO 035568 A1
57. Kaizerman JA, Aaron W, An S, Austin RJ, Brown M, Chong A, Huang T, Hungate R, Jiang B, Johnson MG, Lee G, Lucas BS, Orf J, Rong M, Toteva MM, Wickramasinghe D, Xu G, Ye Q, Zhong W, McMinn DL (2010) Addressing PXR liability of phthalazine-based hedgehog/smoothened antagonists using novel pyridopyridazines. Bioorg Med Chem Lett 20:4607–4610
58. Brown ML, Aaron W, Austin RJ, Chong A, Huang T, Jiang B, Kaizerman JA, Lee G, Lucas BS, McMinn DL, Orf J, Rong M, Toteva MM, Xu G, Ye Q, Zhong W, DeGraffenreid MR, Wickramasinghe D, Powers JP, Hungate R, Johnson MG (2011) Discovery of amide replacement that improve activity and metabolic stability of a bis-amide smoothened antagonist hit. Bioorg Med Chem Lett 21:5206–5209
59. Miller-Moslin K, Peukert S, Jain RK, McEwan MA, Karki R, Llamas L, Yusuff N, He F, Li Y, Sun Y, Dai M, Perez L, Michael W, Sheng T, Lei H, Zhang R, Williams J, Bourret A, Ramamurthy A, Yuan J, Guo R, Matsumoto M, Vattay A, Maniara W, Amaral A, Dorsch M, Kelleher JF 3rd (2009) 1-Amino-4-benzylphthalazines as orally bioavailable smoothened antagonists with antitumor activity. J Med Chem 52:3954–3968
60. Peukert S, He F, Dai M, Zhang R, Sun Y, Miller-Moslin K, McEwan M, Lagu B, Wang K, Yusuff N, Bourret A, Ramamurthy A, Maniara W, Amaral A, Vattay A, Wang A, Guo R, Yuan J, Green J, Williams J, Buonamici S, Kelleher JF 3rd, Dorsch M (2013) ChemMedChem 8:1261–1265
61. Pan S, Wu X, Jiang J, Gao W, Wan Y, Cheng D, Han D, Liu J, Englund NP, Wan Y, Peukert S, Miller-Moslin K, Yuan J, Guo R, Matsumoro M, Vattay A, Jiang Y, Tsao J, Sun F, Pferdekamper AC, Dodd S, Tuntland T, Maniara W, Kelleher 3rd JF, Yao Y, Warmuth M, Williams J, Dorsch M (2010) Discovery of NVP-LDE225, a potent and selective smoothened antagonist. ACS Med Chem Lett 1:130–134
62. Buonamici S, Williams J, Morrissey M, Wang A, Guo R, Vattay A, Hsiao K, Yuan J, Green J, Ospina B, Yu Q, Ostrom L, Fordjour P, Anderson DL, Monahan JE, Kelleher JF, Peukert S, Pan S, Wu X, Maira SM, García-Echeverría C, Briggs KJ, Watkins DN, Yao YM, Lengauer C, Warmuth M, Sellers WR, Dorsch M (2010) Interfering with resistance to smoothened antagonists by inhibition of the PI3K pathway in medulloblastoma. Sci Transl Med 2:51ra70
63. Cheng D, Han D, Gao W, Jing Q, Jiang J, Wan Y, Englund NP, Tuntland T, Wu X, Pan S (2012) Design, synthesis, and structure-activity-relationship of phenyl imidazoles as potent smoothened antagonists. Bioorg Med Chem Lett 22:6573–6576

64. Peukert S, Jain RK, Geisser A, Sun Y, Zhang R, Bourret A, Carlson A, DaSilva J, Ramamurthy A, Kelleher JF (2009) Identification and structure–activity relationships of ortho-biphenyl carboxamides as potent smoothened antagonists inhibiting the Hedgehog signaling pathway. Bioorg Med Chem Lett 19:328–331
65. Hipskind PA, Takakuwa T (2009) Disubstituted phthalazine Hedgehog pathway antagonists. WO 134574 A2
66. Bender MH, Hipskind PA, Capen AR, Cockman M, Credille KM, Gao H, Bastina JA, Clay JM, Lobb KL, Sall DJ, Thompson ML, Wilson T, Wishart GN, Patel BKR (2011) Identification and characterization of a novel smoothened antagonist for the treatment of cancer with deregulated hedgehog signaling. In: Proceedings of the 102nd annual meeting of the American Association for Cancer Research; 2011 April 2–6; Orlando, FL. Philadelphia (PA): AACR; Cancer Res 2011, 71, Abstract number 2819
67. Wang C, Wu X, Katritch V, Han GW, Huang X-P, Liu W, Siu FY, Roth BL, Cherezov V, Stevens RC (2013) Structure of the human smoothened receptor bound to an antitumour agent. Nature 497:338–343
68. http://www.lillyoncologypipeline.com/Pages/hedgehog-smo-antagonist.aspx. Accessed February 2014
69. Bahceci S, Bajjalieh W, Chen J, Epshteyn S, Forsyth TP, Huynh TP, Kim BG, Leahy JW, Lee MS, Lewis GL, Mac MB, Mann G, Marlowe CK, Ridgway BH, Sangalang JC, Shi X, Rakeuchi CS, Wang Y (2008) Inhibitors of the Hedgehog pathway. WO 112913 A1
70. Rocconi RP, Samant L (2013) Methods and compositions for the treatment of ovarian cancer. WO 043255 A1
71. http://www.exelixis.com/pipeline/xl139. Accessed 26 February 2014
72. Strand MF, Wilson SR, Dembinski JL, Holsworth DD, Khvat A, Okun I, Petersen D, Krauss S (2011) A novel synthetic smoothened antagonist transiently inhibits pancreatic adenocarcinoma xenografts in a mouse model. PLoS One 6:e19904
73. Ohashi T, Oguro Y, Tanaka T, Shiokawa Z, Shibata S, Sato Y, Yamakawa H, Hattori H, Yamamoto Y, Kondo S, Miyamoto M, Tojo H, Baba A, Sasaki S (2012) Discovery of pyrrolo [3,2-*c*]quinoline-4-one derivatives as novel hedgehog signaling inhibitors. Bioorg Med Chem 20:5496–5506
74. Ohashi T, Oguro Y, Tanaka T, Shiokawa Z, Tanaka Y, Shibata S, Sato Y, Yamakawa H, Hattori H, Yamamoto Y, Kondo S, Miyamoto M, Nishihara M, Ishimura Y, Tojo H, Baba A, Sasaki S (2012) Discovery of the investigational drug TAK-441, a pyrrolo[3,2-*c*]pyridine derivative, as a highly potent and oral active hedgehog signaling inhibitor: Modification of the core skeleton for improved solubility. Bioorg Med Chem 20:5507–5517
75. Tojo H, Shibata S, Satoh Y, Kawamura M, Inazuka M, Yamakawa H, Kashiwagi M, Miyamoto M, Kondo S, Oohashi T, Oguro Y, Sasaki S (2011) TAK-441, a novel investigational small molecule hedgehog pathway inhibitor for use in cancer therapy. In: Proceedings of the 102nd annual meeting of the American Association for Cancer Research; 2011 Aprile 2–6; Orlando, FL. Philadelphia (PA): AACR; Cancer Res 2011, 71, Abstract number 2823
76. Kogame A, Tagawa Y, Shibata S, Tojo H, Miyamoto M, Tohyama K, Kondo T, Prakash S, Shyu WC, Asahi S (2013) Pharmacokinetic and pharmacodynamic modeling of Hedgehog inhibitor TAK-441 for the inhibition of Gli1 messenger RNA expression and antitumor efficacy in xenografted tumor model mice. Durg Metab Disp 41:727–734
77. Li JJ, Shanmugasundaram V, Reddy S, Fleischer LL, Wang Z, Smith Y, Harter WG, Yue W-S, Swaroop M, Li L, Ji CX, Dettling D, Osak B, Fitzgerald LR, Conradi R (2010) Smoothened antagonists for hair inhibition. Bioorg Med Chem Lett 20:4932–4935
78. Beachy PA, Chen JK, Taipale AJN (2010) Modulators of Hedgehog signaling pathways, compositions and uses related thereto. US 069386 A1
79. Munchhof MJ, Li Q, Shavnya A, Borzillo GV, Boyden TL, Jones CS, LaGreca SD, Martinez-Alsina L, Patel N, Pelletier K, Reiter LA, Robbins MD, Tkalcevic GT (2012) Discovery of PF-04449913, a potent and orally bioavailable inhibitor of smoothened. ACS Med Chem Lett 3:106–111

80. Jackson-Fisher AJ, McMahon MJ, Lam J, Li C, Engstrom LD, Tsaparikos K, Shields DJ, Fang DD, Lira ME, Zhu Z, Robbins MD, Schwab R, Munchhof MJ, VanArsdale T (2011) PF-04449913, a small molecule inhibitor of Hedgehog signaling, is effective in inhibiting tumor growth in preclinical models. In: Proceedings of the 102nd annual meeting of the American Association for Cancer Research; 2011 April 2–6; Orlando, FL. Philadelphia (PA): AACR; Cancer Res 2011, 71, Abstract number 4504
81. Rohner A, Spilker ME, Lam JL, Pascual B, Bartkowski D, Li QJ, Yang AH, Stevens G, Xu M, Wells PA, Planken S, Nair S, Sun S (2012) Effective targeting of Hedgehog signaling in a medulloblastoma model with PF-5274857, a potent and selective smoothened antagonist that penetrates the blood–brain barrier. Mol Cancer Ther 11:57–65
82. Yang H, Xiang J, Wang N, Zhao Y, Hyman J, Li S, Jiang J, Chen JK, Yang Z, Lin S (2009) Converse conformational control of smoothened activity by structurally related small molecules. J Biol Chem 284:20876–20884
83. Wilson CW, Chen M-H, Chuang P-T (2009) Smoothened adopts multiple active and inactive conformations capable of trafficking to the primary cilium. PloS One 4:e5182
84. Rohatgi R, Milenkovic L, Corcoran RB, Scott MP (2009) Hedgehog signal transduction by smoothened: pharmacologic evidence for a 2-step activation process. Proc Natl Acad Sci USA 106:3196–3201
85. Weierstall U, James D, Wang C, White TA, Wang D, Liu W, Spence JC, Bruce Doak R, Nelson G, Fromme P, Fromme R, Grotjohann I, Kupitz C, Zatsepin NA, Liu H, Basu S, Wacker D, Han GW, Katritch V, Boutet S, Messerschmidt M, Williams GJ, Koglin JE, Marvin Seibert M, Klinker M, Gati C, Shoeman RL, Barty A, Chapman HN, Kirian RA, Beyerlein KR, Stevens RC, Li D, Shah ST, Howe N, Caffrey M, Cherezov V (2014) Lipidic cubic phase injector facilitates membrane protein serial femtosecond crystallography. Nat Commun 5:3309–3309
86. Che C, Li S, Yang B, Xin S, Yu Z, Shao T, Tao C, Lin S, Yang Z (2012) Synthesis and characterization of Sant-75 derivatives as Hedgehog-pathway inhibitors. Beilstein J Org Chem 8:841–849
87. Manetti F, Faure H, Roudaut H, Gorojankina T, Traiffort E, Schoenfelder A, Mann A, Solinas A, Taddei M, Ruat M (2010) Virtual screening-based discovery and mechanistic characterization of the acylthiourea MRT-10 family as smoothened antagonists. Mol Pharmacol 78:658–665
88. Roudaut H, Traiffort E, Gorojankina T, Vincent L, Faure H, Schoenfelder A, Mann A, Manetti F, Solinas A, Taddei M, Ruat M (2011) Identification and mechanism of action of the acylguanidine MRT-83, a novel potent smoothened antagonist. Mol Pharmacol 79:453–460
89. Solinas A, Faure H, Roudaut H, Traiffort E, Schoenfelder A, Mann A, Manetti F, Taddei M, Ruat M (2012) Acylthiourea, acylurea, and acylguanidine derivatives with potent hedgehog inhibiting activity. J Med Chem 55:1559–1571
90. Lipinski RJ, Bushman W (2010) Identification of Hedgehog signaling inhibitors with relevant human exposure by small molecule screening. Toxicol In Vitro 24:1404–1409
91. Wang J, Lu J, Mook RA Jr, Zhang M, Zhao S, Barak LS, Freedman JH, Lyerly HK, Chen W (2012) The insecticide synergist piperonyl butoxide inhibits Hedgehog signaling: assessing chemical risks. Toxicol Sci 128:517–523
92. Wang J, Mook RA Jr, Lu J, Gooden DM, Ribeiro A, Guo A, Barak LS, Lyerly HK, Chen W (2012) Identification of a novel smoothened antagonist that potently suppresses Hedgehog signaling. Bioorg Med Chem 20:6751–6757
93. Chen JK, Taipale J, Young KE, Maiti T, Beachy PA (2002) Small molecule modulation of smoothened activity. Proc Natl Acad Sci USA 99:14071–14076
94. Rominger CM, Bee W-LT, Copeland RA, Davenport EA, Gilmartin A, Gontarek R, Hornberger KR, Kallal LA, Lai Z, Lawrie K, Lu Q, McMillan L, Truong M, Tummino PJ, Turunen B, Will M, Zuercher WJ, Rominger DH (2009) Evidence for allosteric interactions of antagonist binding to the smoothened receptor. J Pharm Exp Ther 329:995–1005
95. Chen JK, Sinha S (2006) Purmorphamine activates the Hedgehog pathway by targeting smoothened. Nat Chem Biol 2:29–30

96. Buttner A, Seifert K, Cottin T, Sarli V, Tzagkaroulaki L, Scholz S, Giannis A (2009) Synthesis and biological evaluation of SANT-2 and analogues as inhibitors of the Hedgehog signaling pathway. Bioorg Med Chem 17:4943–4954
97. Zhao Y, Tong C, Jiang J (2007) Hedgehog regulates smoothened activity by inducing a conformational switch. Nature 450:252–258
98. Tao H, Jin Q, Koo D-I, Liao X, Englund NP, Wang Y, Ramamurthy A, Schultz PG, Dorsch M, Kelleher J, Wu X (2011) Small molecule antagonists in distinct binding modes inhibit drug-resistant mutant of smoothened. Chem Biol 18:432–437
99. Wang Y, Arvanites AC, Davidow L, Blanchard J, Lam K, Yoo JW, Coy S, Rubin LL, McMahon AP (2012) Selective identification of Hedgehog pathway antagonists by direct analysis of smoothened ciliary translocation. ACS Chem Biol 7:1040–1048
100. Hyman JM, Firestone AJ, Heine VM, Zhao Y, Ocasio CA, Han K, Sun M, Rack PG, Sinha S, Wu JJ, Solow-Cordero DE, Jiang J, Rowitch DH, Chen JK (2009) Small-molecule inhibitors reveal multiple strategies for Hedgehog pathway blockade. Proc Natl Acad Sci USA 106:14132–14137
101. Kim J, Tang JY, Gong R, Kim J, Lee JJ, Clemons KV, Chong CR, Chang KS, Fereshteh M, Gardner D, Reya T, Liu JO, Epstein EH, Stevens DA, Beachy PA (2010) Itraconazole, a commonly used antifungal that inhibits Hedgehog pathway activity and cancer growth. Cancer Cell 17:388–399
102. Kim J, Lee JJ, Kim J, Gardner D, Beachy PA (2010) Arsenic antagonizes the Hedgehog pathway by preventing ciliary accumulation and reducing stability of the Gli2 transcriptional effector. Proc Natl Acad Sci USA 107:13432–13437
103. Beauchamp EM, Ringer L, Bulut G, Sajwan KPM, Hall MD, Lee Y-C, Peaceman D, Ozdemirli M, Rodriguez O, Macdonald TJ, Albanese C, Toretsky JA, Uren A (2011) Arsenic trioxide inhibits human cancer cell growth and tumor development in mice by blocking Hedgehog/GLI pathway. J Clin Invest 121:148–160
104. Kim J, Aftab BT, Tang JY, Kim D, Lee AH, Rezaee M, Kim J, Chen B, King EM, Borodovsky A, Riggins GJ, Epstein EH Jr, Beachy PA, Rudin CM (2013) Itraconazole and arsenic trioxide inhibit Hedgehog pathway activation and tumor growth associated with acquired resistance to smoothened antagonists. Cancer Cell 23:23–34
105. Nedelcu D, Liu J, Xu Y, Jao C, Salic A (2013) Oxysterol binding to the extracellular domain of smoothened in Hedgehog signaling. Nat Chem Biol 9:557–564
106. Rana R, Carroll CE, Lee H-J, Bao J, Marada S, Grace CRR, Guibao CD, Ogden SK, Zheng JJ (2013) Structural insights into the role of the smoothened cysteine-rich domain in Hedgehog signalling. Nat Commun 4:2965
107. Wang Y, Davidow L, Arvanites AC, Blanchard J, Lam K, Xu K, Oza V, Yoo JW, Ng JMY, Curran T, Rubin LL, McMahon AP (2012) Glucocorticoid compounds modify smoothened localization and Hedgehog pathway activity. Chem Biol 19:972–982
108. Yang B, Hird AW, Russell DJ, Fauber BP, Dakin LA, Zheng X, Su Q, Godin R, Brassil P, Devereaux E, Janetka JW (2012) Discovery of novel Hedgehog antagonists from cell-based screening: isosteric modification of p38 bisamides as potent inhibitors of SMO. Bioorg Med Chem Lett 22:4907–4911
109. Xin M, Wen J, Tang F, Tu C, Shen H, Zhao X (2013) The discovery of novel N-(2-pyrimidinylamino)benzamide derivatives as potent Hedgehog signaling pathway inhibitors. Bioorg Med Chem Lett 23:6777–6783
110. Xin M, Wen J, Tang F, Tu C, Huang W, Shen H, Zhao X, Cheng L, Wang M, Zhang L (2014) Synthesis and evaluation of 4-(2-pyrimidinylamino)benzamides inhibitors of hedgehog signaling pathway. Bioorg Med Chem Lett 24:983–988
111. Xin M, Zhang L, Tang F, Tu C, Wen J, Zhao X, Liu Z, Cheng L, Shen H (2014) Design, synthesis, and evaluation of pyrrolo[2,1-*f*][1,2,4]triazine derivatives as novel hedgehog signaling pathway inhibitors. Bioorg Med Chem 22:1429–1440
112. Stanton BZ, Peng LF, Maloof N, Nakai K, Wang X, Herlihy KM, Duffner JL, Taveras KM, Hyman JM, Lee SW, Koehler AN, Chen JK, Fox JL, Mandinova A, Schreiber SL (2009) A

small molecule that binds Hedgehog and blocks its signaling in human cells. Nat Chem Biol 5:154–156

113. Peng LF, Stanton BZ, Maloof N, Wang X, Schreiber SL (2009) Syntheses of aminoalcohol-derivative macrocycles leading to a small-molecule binder to and inhibitor of Sonic Hedgehog. Bioorg Med Chem Lett 19:6319–6325
114. Dockendorff C, Nagiec MM, Weiwer M, Buhrlage S, Ting A, Nag PP, Germain A, Kim H-J, Youngsaye W, Scherer C, Bennion M, Xue L, Stanton BZ, Lewis TA, MacPherson L, Palmer M, Foley MA, Perez JR, Schreiber SL (2012) Macrocyclic Hedgehog pathway inhibitors: optimization of cellular activity and mode of action studies. ACS Med Chem Lett 3:808–813
115. Schaefer GI, Perez JR, Duvall JR, Stanton BZ, Shamji AF, Schreiber SL (2013) Discovery of small-molecule modulators of the Sonic Hedgehog pathway. J Am Chem Soc 135:9675–9680
116. Narayan R, Bauer JO, Strohmann C, Antonchick AP, Waldmann H (2013) Catalytic enantioselective synthesis of functionalized tropanes reveals novel inhibitors of Hedgehog signaling. Angew Chem Int Ed 12892–12896
117. Arai MA, Tateno C, Hosoya T, Koyano T, Kowithayakorn T, Ishibashi M (2008) Hedgehog/GLI-mediated transcriptional inhibitors from *Zizyphus cambodiana*. Bioorg Med Chem 16:9420–9424
118. Arai MA, Tateno C, Koyano T, Kowithayakorn T, Kawabe S, Ishibashi M (2011) New Hedgehog/GLI-signaling inhibitors from *Adenium obesum*. Org Biomol Chem 9:1133–1139
119. Rifai Y, Arai MA, Koyano T, Kowithayakorn T, Ishibashi M (2011) Acoschimperoside P, 2′ acetate: a Hedgehog signaling inhibitory constituent from *Vallaris glabra*. J Nat Med 65:629–632
120. Lu Y, Lin H, Drenkhahn SK, Starkey NJE, Jackson GA, Folk W, Fritsche KL, Lubahn DB (2011) Inhibition of Hedgehog signaling by extract of Sutherlandia. In: Proceedings of the 102nd annual meeting of the American Association for Cancer Research; 2011 April 2–6; Orlando, FL. Philadelphia (PA): AACR; Cancer Res 2011, 71, Abstract number 4221
121. Rifai Y, Arai MA, Koyano T, Kowithayakorn T, Ishibashi M (2010) Terpenoids and a flavonoid glycoside from *Acacia pennata* leaves as Hedgehog/GLI-mediated transcriptional inhibitors. J Nat Prod 73:995–997
122. Slusarz A, Shenouda NS, Sakla MS, Drenkhahn SK, Narula AS, MacDonald RS, Besch-Williford CL, Lubahn DB (2010) Common botanical compounds inhibit the Hedgehog signaling pathway in prostate cancer. Cancer Res 70:3382–3390
123. Hosoya T, Arai MA, Koyano T, Kowithayakorn T, Ishibashi M (2008) Naturally occurring small-molecule inhibitors of hedgehog/GLI-mediated transcription. Chembiochem 9:1082–1092
124. Shintani A, Toume K, Rifai T, Arai MA, Ishibashi M (2010) A bisindole alkaloid with Hedgehog signal inhibitory activity from the Myxomycete *Perichaena chrysosperma*. J Nat Prod 73:1711–1713
125. Mahindroo N, Connelly MC, Punchihewa C, Kimura H, Smeltzer MP, Wu S, Fujii N (2009) Structure–activity relationships and cancer-cell selective toxicity of novel inhibitors of glioma-associated oncogene homologe 1 (Gli1) mediated transcription. J Med Chem 52:4277–4287
126. He B, Fujii N, You L, Xu Z, Jablons DM (2007) Targeting Gli proteins in human cancer by small molecules. WO 067814 A2
127. Mahindroo N, Connelly MC, Punchihewa C, Yang L, Yan B, Fujii N (2010) Amide conjugates of ketoprofen and indole as inhibitors of Gli1-mediated transcription in the Hedgehog pathway. Bioorg Med Chem 18:4801–4811
128. Lauth M, Bergstrom A, Shimokawa T, Toftgard R (2007) Inhibition of GLI-mediated transcription and tumor cell growth by small-molecule antagonists. Proc Natl Acad Sci USA 104:8455–8460
129. Hayakawa I, Agatsuma T, Furukawa H, Kurakata S, Naruto S (2002) Imidazopyridine derivatives. WO 034748 A1

130. Lee J, Wu X, Pasca di Magliano M, Peters EC, Wang Y, Hong J, Hebrok M, Ding S, Cho CY, Schultz PG (2007) A small-molecule antagonist of the Hedgehog signaling pathway. ChemBioChem 8:1916–1919
131. Cupido T, Rack PG, Firestone AJ, Hyman JM, Han K, Sinha S, Ocasio CA, Chen JK (2009) The imidazopyridine derivative JK184 reveals dual roles for microtubules in hedgehog signaling. Angew Chem Int Ed 48:2321–2324
132. Beachy PA, Chen JK, Taipale J (2005) Hedgehog pathway antagonists. WO 033288 A2
133. Firestone AJ, Weinger JS, Maldonado M, Barlan K, Langston LD, O'Donnell M, Gelfand VI, Kapoor TM, Chen JK (2012) Small-molecule inhibitors of the AAA + ATPase motor cytoplasmic dynein. Nature 484:125–129
134. Baxter AD, Boyd EA, Guicherit OM, Porter J, Price S, Rubin LE (2001) Small organic molecule regulators of cell proliferation. WO 01/74344 A2
135. King RW (2002) Roughing up smoothened: chemical modulators of Hedgehog signaling. J Biol 1:8
136. Brunton SA, Stibbard JHA, Rubin LL, Guicherit OM, Kruse LI, Price S, di Lucrezia R, MacKinnon CH, Avery A, Park I, Buxton D, Boyd EA (2009) Potent agonists of the Hedgehog signaling pathway. Bioorg Med Chem Lett 19:4308–4311
137. Seifert K, Buttner A, Rigol S, Eilert N, Wandel E, Giannis A (2012) Potent small molecule Hedgehog agonists induce VEGF expression in vitro. Bioorg Med Chem 20:6465–6481
138. Wu X, Ding S, Ding Q, Gray NS, Schultz PG (2002) A small molecule with osteogenesis-inducing activity in multipotent mesenchymal progenitor cells. J Am Chem Soc 124:14520–14521
139. Lee S-J, Lee H-K, Cho SY, Choi J-K, Shin HK, Kwak E-J, Cho M-R, Kim H-R, Kim S-R, Kim Y-M, Park K-J, Choi J-K (2008) Identification of osteogenic purmorphamine derivatives. Mol Cells 26:380–386
140. Chen S, Zhang Q, Wu X, Schultz PG, Ding S (2004) Dedifferentiation of lineage-committed cells by a small molecule. J Am Chem Soc 126:410–411
141. Chen W, Barak LS, Lyerly HK, Wang J, Bond MC, Mook RA Jr (2011) Glucocorticoid drugs as smoothened agonists. WO 109711 A1
142. Wang J, Lu J, Bond MC, Chen M, Ren X-R, Lyerly HK, Barak LS, Chen W (2010) Identification of select glucocorticoids as smoothened agonists: potential utility for regenerative medicine. Proc Natl Acad Sci USA 107:9323–9328
143. Nachtergaele S, Mydock LK, Krishnan K, Rammohan J, Schlesinger PH, Covey DF, Rohatgi R (2012) Oxysterols are allosteric activators of the oncoprotein smoothened. Nat Chem Biol 8:211–220
144. Montgomery SR, Nargizyan T, Meliton V, Nachtergaele S, Rohatgi R, Stappenbeck F, Jung ME, Johnson JS, Aghdasi B, Tian H, Weintraub G, Inoue H, Atti E, Tetradis S, Pereira RC, Hokugo A, Alobaidaan R, Tan Y, Hahn TJ, Wang JC, Parhami F (2014) A novel osteogenic oxysterol compound for therapeutic development to promote bone growth: activation of Hedgehog signaling and osteogenesis through smoothened binding. J Bone Miner Res 29:1872–1885
145. Parhami F, Jung ME, Stappenbeck F, Pierce WM, Taylor GK, Merten KE (2013) Novel oxysterol analogue, Oxy149, induces osteogenesis and Hedgehog signaling and inhibits adipogenesis. WO 169397 A1
146. Corman A, DeBerardinis AM, Hadden MK (2012) Structure–activity relationships for side chain oxysterol agonists of the Hedgehog signaling pathway. ACS Med Chem Lett 3:828–833
147. Gorojankina T, Hoch L, Faure H, Roudaut H, Traiffort E, Schoenfelder A, Girard N, Mann A, Manetti F, Solinas A, Petricci E, Taddei M, Ruat M (2013) Discovery, molecular and pharmacological characterization of GSA-10, a novel small molecule positive modulator of smoothened. Mol Pharmacol 83:1020–1029

Top Med Chem (2015) 16: 109–126
DOI: 10.1007/7355_2014_62

Published online: 25 October 2014

Receptor Modifications in Hedgehog Regulation

Qing Shi and Jin Jiang

Abstract The Hedgehog (Hh) signaling pathway is one of the highly conserved signaling cascades that control cell growth, cell fate, and pattern formation in species ranging from *Drosophila* to human. Smoothened (Smo), a G-protein-coupled-receptor (GPCR) family protein, serves as a core component to tranduce Hh signal across the cell membrane. Studying how the activity of Smo is regulated is a key to understand how Hh morphegen gradiant differentially induces target gene expression during normal animal development as well as how abnormal Hh signaling activity contributes to various human diseases. In this chapter, we focus on the regulation of Smo by its posttranslational modifications. By reviewing our current knowledge on how Hh ligands dynamically induce phosphorylation and ubiquitination of Smo and how these modifications coordinatedly modulate Smo function and downstream signaling events, we hope to inspire future exploration of comprehensive mechanisms underlying the regulation of Smo.

Keywords Hedgehog, Phospohrylation, Posttranslational modification, Smo, Smo binding proteins, Ubiquitination

Q. Shi
Department of Developmental Biology, University of Texas Southwestern Medical Center at Dallas, Dallas, TX 75390, USA

J. Jiang (✉)
Department of Developmental Biology, University of Texas Southwestern Medical Center at Dallas, Dallas, TX 75390, USA

Department of pharmacology, University of Texas Southwestern Medical Center at Dallas, Dallas, TX 75390, USA
e-mail: jin.jiang@utsouthwestern.edu

Contents

Abbreviations

A	Anterior
BCC	Basal cell carcinoma
CFP	Cyan fluorescent protein
Ci^A	Ci activator
Ci^R	Ci repressor
CK1	Casein kinase 1
CK2	Casein kinase 2
Cos2	Costal2
C-tail	Carboxyl-terminal cytoplasmic tail
dSmo	*Drosophila* Smo
DUBs	Deubiquitinating enzymes
E1	An Ub-activation enzyme
E2	An Ub-conjugating enzyme
E3	An Ub ligase
FRET	Fluorescence resonance energy transfer
Fu	Fused
GPCR	G-protein-coupled-receptor
Gprk2/GRK2	G-protein-coupled-receptor kinase 2
GSK3	Glycogen synthase kinase 3
Hh	Hedgehog
Krz	Kurtz
L3	The third intracellular loop
mSmo	Mammalian Smo
P	Posterior
PKA	Protein kinase A
PP1	Protein phosphatase 1
PP2A	Protein phosphatase 2A
Ptc	Patched
PTM	Posttranslational modification

RA	Arg to Ala
SAID	Smo auto-inhibitory domain
Ub	Ubiquitin
YFP	Yellow fluorescent protein

1 Introduction

Posttranslational modification (PTM) is a critical step in protein biosynthesis that increases the functional diversity of the proteome by the covalent addition of functional chemical groups to proteins. Within the last few decades, numerous types of PTMs, such as phosphorylation, glycosylation, ubiquitination, nitrosylation, methylation, etc., have been uncovered to influence almost all aspects of normal cell biology and pathogenesis. It is thus not surprising that components of Hedgehog (Hh) pathway are dynamically modulated by various PTMs. For example, dual lipid modifications facilitate multimerization of Hh and are essential for its ligand activity [1–3]. Ubiquitination controls the turnover of Patched (Ptc) [4], which is a twelve transmembrane protein serving as the Hh receptor [5]. Phosphorylation coupled with ubiquitination imposes multiple layers of regulation on the Hh pathway components including the transcription factor, Ci/Gli [6–9]. In this chapter, we will review the PTMs involved in the regulation of Smo, a G-protein-coupled-receptor (GPCR)-like protein, which functions as an obligated signal transducer of the canonical Hh signaling across the plasma membrane. We will begin with a brief overview of the role of Smo in the Hh signal transduction. We will then review two types of PTMs, phosphorylation and ubiquitination, in the regulation of Smo by discussing how these PTMs are dynamically regulated by the Hh ligand, how they regulate Smo function, and how they coordinate with Smo binding proteins to transduce the Hh signal. Finally, we will highlight the unresolved questions regarding the regulation of Smo by PTMs and its interacting proteins for future investigation.

2 Smo Transduces Hh Signal Across the Plasma Membrane

The Hh signaling pathway is one of the major developmental pathways that control cell growth, cell fate, and pattern formation in species ranging from *Drosophila* to human [7, 10]. *smo*, first identified as a segment polarity gene in *Drosophila* by Nusslein-Volhard and Wieschaus [11], encodes a conserved GPCR-like protein that serves as a core component of the Hh reception system across the cell membrane [12, 13]. As gain-of-function mutations in Smo contribute to various human cancers, including basal cell carcinoma (BCC), the most common type of cancer afflicting ~750,000 people each year in the USA alone, and medulloblastoma, the most common childhood brain cancer [14–16], Smo has emerged as a prominent target for drug development and cancer therapeutics [17].

In the absence of the Hh ligand, Smo is inhibited by the Hh receptor Ptc through an ill-defined mechanism, likely through the transporter activity of Ptc [7, 10, 18, 19]. In *Drosophila*, Ptc restricts Smo cell surface expression by promoting its endocytosis and degradation [20–22], whereas in mammals, Ptc prevents Smo from accumulating in the primary cilium, a specific microtubule-based cell surface protrusion essential for mammalian Hh signal transduction [23, 24]. Binding of the Hh ligand to Ptc alleviates its inhibition on Smo by reciprocally regulating the subcellular distribution of Ptc and Smo (Fig. 1). In *Drosophila*, Hh promotes Ptc endocytosis but stimulates Smo accumulation on the cell surface [20, 21, 25], whereas in mammals, Hh reception triggers Ptc moving out of but Smo accumulating in the primary cilium [23, 24]. The enrichment of Smo either on the cell surface or in the primary cilium correlates with Hh pathway activation, but the underlying signal transduction mechanisms remain poorly understood, especially in mammals.

Upon activation, Smo relays the Hh signal to the transcription factor Ci/Gli via multi-protein signaling complexes (Fig. 1). In *Drosophila*, a kinesin-like protein Costal2 (Cos2) serves as a molecular scaffold to bring together Ci and several kinases, including the serine threonine kinase Fused (Fu), Protein kinase A (PKA), Glycogen synthase kinase 3 (GSK3), and Casein kinase 1 (CK1), leading to efficient phosphorylation and proteolytic processing of Ci into a truncated repressor form (Ci^R) in the absence of Hh [26–30]. Hh-induced Smo activation blocks Ci^R production by dissociating the Cos2-Ci-kinase complexes [29, 30], and converts the full-length Ci into a hyperactive form (Ci^A) by recruiting and activating the Cos2/Fu complex to release Sufu-mediated suppression of Ci [31, 32]. In vertebrates, the Cos2 homolog Kif7 also plays a dual role in the Hh signaling [33–37]. On the one hand, Kif7 forms a complex with Gli transcription factors and its inactivation leads to defective Gli processing and thus ectopic Hh pathway activation. On the other hand, Kif7 is also required for maximal Hh pathway activation likely by interacting with the activated Smo. Despite the conserved role of Cos2/Kif7 in the Hh signaling pathway, the pathway components downstream of Smo diverges significantly between flies and mammals. For example, the mammalian homolog of Fu kinase is dispensable for Hh signaling [38, 39], and hence other kinase(s) may function as a substitute for Fu in the regulation of Gli activity [40, 41]. In *Drosophila*, $G\alpha i$ has been implicated acting downstream of Smo and is associated with Cos2 in response to Hh stimulation [42]; however, a role for $G\alpha i$ in mammalian Hh signaling has remained controversial [43].

How Hh and Ptc reciprocally regulate Smo intracellular/ciliary trafficking and activity, and how Smo transduces the Hh signal across the plasma membrane to activate the intracellular signaling cascade are fundamental questions in the field. In the following chapter, we will review the progress made in the past decade toward addressing these important questions, with a focus on Smo PTM and its role in the Hh signal transduction.

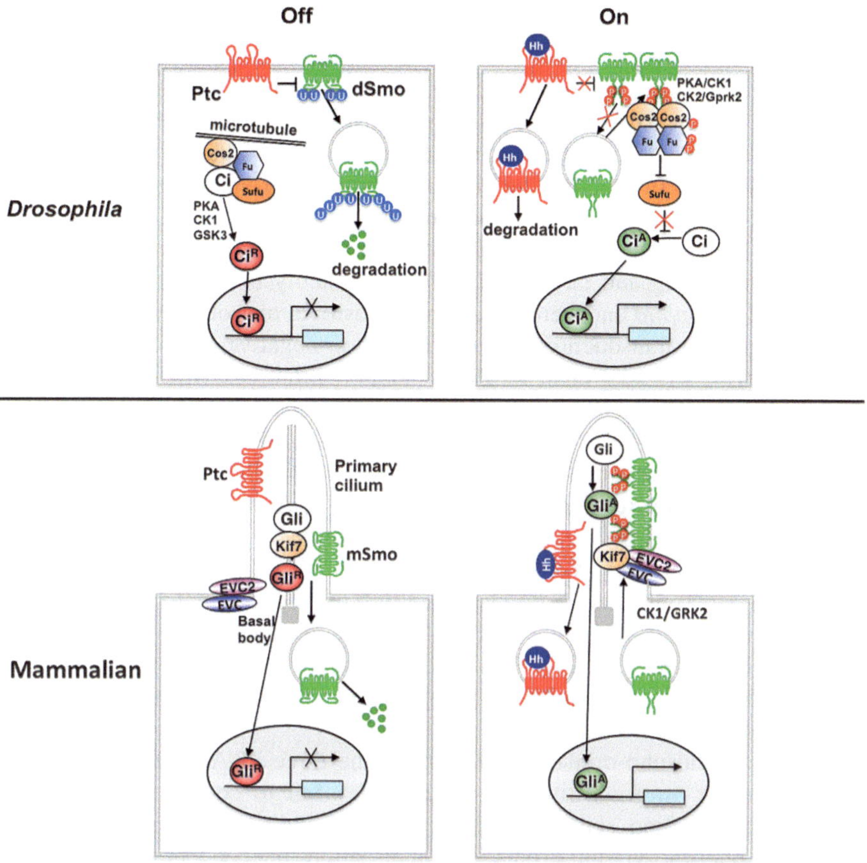

Fig. 1 Comparison of *Drosophila* and mammalian Hh signal pathways. In the absence of Hh ligand ("Off" state), Ptc inhibits Smo phosphorylation and keeps Smo in a "closed" conformation. In *Drosophila*, Smo is ubiquitinated at multiple sites, which drives Smo internalization and degradation whereas in mammals, Smo is excluded from the primary cilia where Ptc resides. Ci/Gli is phosphorylated by multiple kinases, which targets it for Slimb/βTRCP-mediated proteolysis to generate the truncated repressor form Ci^R/Gli^R. In the presence of Hh ligand ("On" state), binding of Hh to Ptc releases its inhibition on Smo and triggers phosphorylation of Smo by multiple kinases (PKA, CK1, and Gprk2 in *Drosophila*, and CK1 and GRK2 in mammals), leading to Smo cell surface (in *Drosophila*) or ciliary (in mammals) accumulation and a switch from a "closed" and inactive conformation to an "open" and active conformation. In *Drosophila*, the activated Smo recruits the Cos2-Fu complex to the cell surface and induces Fu dimerization and phosphorylation, leading to its activation. Activated Fu inhibits Ci processing and converts Ci into an activator form Ci^A by antagonizing the inhibition imposed by Sufu. In mammals, activated Smo interacts with Evc/Evc2/Kif7 and converts Gli into its activator form Gli^A in the primary cilium

2.1 Hh Induces Smo Phosphorylation by Multiple Kinases

Phosphorylation is a reversible process in which protein kinases covalently add phosphate groups to serine/threonine or tyrosine residues in proteins to impose regulation on a wide range of cellular processes by altering the activities of the modified proteins, facilitating their intracellular sorting, modulating their interactions, or targeting them for degradation [44–47].

Initial studies identified PKA and CK1 as negative regulators of the Hh pathway by phosphorylating and targeting Ci/Gli for proteolysis [48–52]. However, later studies in *Drosophila* uncovered unexpected positive roles of both PKA and CK1 in the Hh signaling transduction in addition to their negative roles in promoting Ci^R production [9, 21, 53]. Gain of PKA function promotes whereas loss of PKA function blocks Hh-induced Smo accumulation on the cell surface as well as high levels of Hh signaling [21]. Similarly, inactivation of CK1α/ε by RNAi blocks Hh-induced Smo accumulation and high levels of Hh signaling activity [21]. Biochemical studies revealed that PKA and CK1 sequentially phosphorylate three clusters of Ser/Thr residues in the carboxyl-terminal cytoplasmic tail (C-tail) of Smo, with PKA serving as a priming kinase for CK1 phosphorylation (Fig. 2) [21, 54, 55]. Phospho-deficient Smo variants exhibited reduced cell surface expression and diminished Hh signaling activity whereas phospho-mimetic Smo variants exhibited increased cell surface level and ligand-independent constitutive activity [21, 54, 55], suggesting that PKA/CK1-mediated phosphorylation is both necessary and sufficient for Smo activation.

Although phospho-mimetic mutations in the three PKA/CK1 phosphorylation clusters render Smo constitutively active, they fail to confer full activity [25], suggesting that Smo activation likely involves additional mechanisms such as additional phosphorylation events, phosphorylation-independent regulatory events, or both. A Mass Spec analysis identified many phosphorylation sites in the Smo C-tail besides the three PKA/CK1 phosphorylation clusters [55], suggesting that Smo is likely to be regulated by additional kinases. Indeed, genetic modifier screens identified Casein kinase 2 (CK2) and G-protein-coupled-receptor kinase 2 (Gprk2/GRK2) as positive regulators of the Hh pathway [56, 57]. In vitro kinase assay indicated that CK2 directly phosphorylated multiple Ser residues in the Smo C-tail [56], some of which were also phosphorylated in cultured cells in the presence of Hh ligand [55]. Gprk2 phosphorylates Smo C-tail at Ser741 and Thr742, which is facilitated by PKA/CK1-mediated phosphorylation at adjacent Ser residues (Fig. 2) [57]. Interestingly, Gprk2 promotes high-level Hh signaling by regulating the active state of Smo through both kinase-dependent and kinase activity-independent mechanisms [57, 58]. In addition, Gprk2 expression is induced by Hh signaling, which forms a positive feedback loop to facilitate high-level Hh signaling [57, 58].

Mammalian Smo (mSmo) diverges significantly from *Drosophila* Smo (dSmo) in the primary sequence and its C-tail does not contain the three PKA/CK1 phosphorylation clusters found in dSmo C-tail (Fig. 2); however, a recent study revealed that mSmo is activated by multi-site phosphorylation similar to dSmo. Although PKA is not involved in mSmo phosphorylation and activation, both CK1α

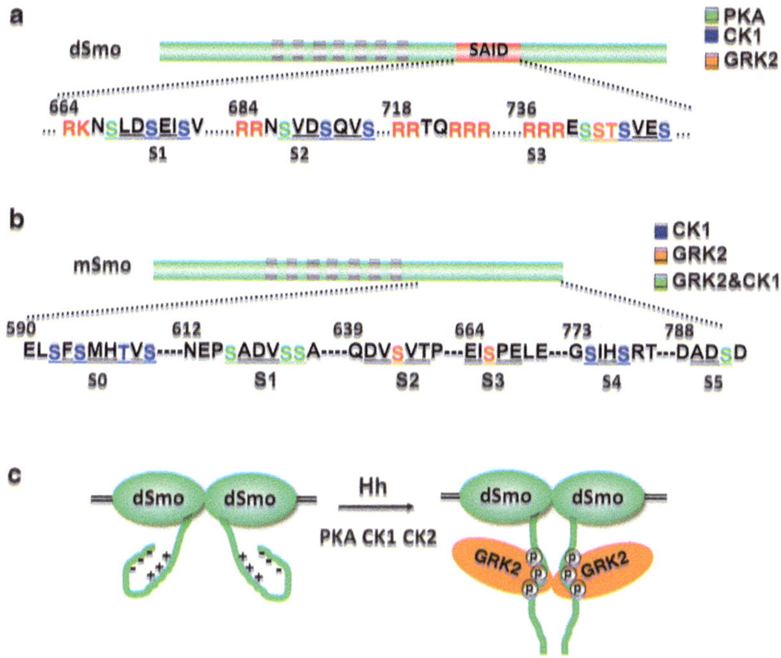

Fig. 2 Both *Drosophila* and mammalian Smo proteins are regulated by multi-site phosphorylation. Diagrams showing *Drosophila* (**a**) and mammalian (**b**) Smo proteins and multiple phosphorylation sites in their C-tails. The Arg motifs adjacent to the phosphorylation clusters in dSmo are highlighted in red. The red box in dSmo C-tail denotes the autoinhibitory domain SAID. (**c**) Diagrams of the "closed" inactive (left) and "open" active (right) conformation of dSmo. The "closed" conformation is maintained by intramolecular electrostatic interactions between multiple Arg motifs in the SAID domain and multiple acidic clusters near the C-terminal region of dSmo. Hh-induced phosphorylation by PKA and CK1 neutralizes the positive charge and thus disrupts the intramolecular electrostatic interactions, leading to unfolding and dimerization/oligomerization of dSmo C-tails. Binding of Gprk2 to dSmo C-tail stabilizes its active conformation and facilitates its dimerization/oligomerization

and GRK2 are required for the Hh-induced mSmo phosphorylation and pathway activation [59–62]. Both CK1α and GRK2 bind mSmo in response to Hh stimulation to phosphorylate mSmo C-tail on at least six clusters of Ser/Thr residues (Fig. 2) [61]. Oncogenic mutations promote mSmo phosphorylation by CK1α and GRK2, and blocking these phosphorylation events inhibits the activity of the oncogenic forms of Smo [61], suggesting that interfering with Smo phosphorylation may represent a new approach to treat cancers caused by *Ptc* or *Smo* mutations.

CK1α/GRK2-mediated phosphorylation promotes mSmo ciliary localization by recruiting β-arrestin, which links mSmo to the anterograde kinesin-II motor [61, 63]. Interestingly, a small molecule, cyclopamine, promotes mSmo ciliary localization but blocks Smo phosphorylation and Hh pathway activation [64–66], suggesting that ciliary localization of mSmo is insufficient for Hh pathway activation and that the Hh-induced mSmo phosphorylation imposes additional layer (s) of regulation leading to Smo activation.

2.2 Phosphorylation Activates Smo by Inducing a Conformational Switch

By employing fluorescence resonance energy transfer (FRET) analysis, Zhao et al. provided evidence for a conformational change of Smo C-tail upon Hh-stimulated Smo phosphorylation [25]. Like other members of the GPCR family, Smo forms a constitutive dimer through interactions mediated by its extracellular domain or/and transmembrane helixes, as high FRET was observed between N-terminally inserted cyan fluorescent protein (CFP) and yellow fluorescent protein (YFP) regardless of the presence or absence of Hh. However, in the absence of Hh, Smo adopts a "closed" conformation in which the Smo C-tail folds back with the C-terminus in close proximity to its third intracellular loop (L3). The closed conformation of Smo C-tail is maintained through electrostatic interactions between multiple clusters of basic residues (mostly Arginine and thus called Arg-motifs) in the Smo auto-inhibitory domain (SAID) and multiple acidic clusters near the Smo C-terminus (Fig. 2). Deletion of the SAID domain or mutating multiple Arg motifs in the SAID domain resulted in constitutive activation of the mutant forms of Smo [25].

Upon Hh stimulation, PKA and CK1 phosphorylate Smo at the three clusters of sites in the SAID domain, which brings in negative charges to neutralize the positive charges of the neighboring Arg motifs and thus disrupts the electrostatic interactions required for maintaining the "closed" conformation, allowing Smo C-tail to unfold and adopt an "open" conformation (Fig. 2). Phosphorylation deficient mutations of the PKA/CK1 sites lock Smo C-tail in its "closed" conformation whereas phospho-mimetic mutations at these sites promote the "open" conformation [25]. In addition, the PKA/CK1-mediated phosphorylation enhances the binding of Gprk2, which phosphorylates Smo at Ser741 and Thr742 to bring in additional negative charges to facilitate the transition of Smo C-tail from the "closed" to the "open" conformation. Physical interaction between Gprk2 and Smo C-tail may also stabilize Smo in its "open" conformation. Furthermore, Gprk2 itself forms a dimer/oligomer, thus the binding of Gprk2 to Smo C-tail facilitates its clustering (Fig. 2) [57].

In mammals, Hh induces a conformational switch of mSmo C-tail similar to that of dSmo [25], which is governed by CK1/GRK2-mediated phosphorylation [61]. A similar conformational change is also induced by the oncogenic Smo mutation (A1/M2) and Smo agonist SAG but blocked by cyclopamine [61]. Cyclopamine traps ciliary localized mSmo in an unphosphorylated/hypophosphorylated form that adopts an inactive conformation whereas ciliary localized mSmo in response to Hh or Smo agonists is phosphorylated and thus adopts an active conformation [61], suggesting that phosphorylation-induced mSmo conformational switch represents an additional and critical step in the activation of mSmo. Hence, Smo phosphorylation provides a more faithful readout for Hh pathway activation than Smo ciliary localization and can serve as a biomarker for cancers caused by deregulated Hh pathway activation.

2.3 Phosphorylation-Induced Smo Conformational Switch Recruits Intracellular Signaling Complexes

Increasing the proximity of Smo C-tails within a Smo dimer/oligomer appears to be crucial for Hh pathway activation because point mutations that impair Smo dimerization compromises Smo activity, which can be partially rescued by forced dimerization through a heterologous system [25]. Furthermore, induced clustering of Smo C-tails suffices to promote Hh pathway activation [25]. How do conformational change and clustering of Smo C-tails lead to the activation of intracellular signaling components? Studies in *Drosophila* S2 cells revealed that Hh promotes colocalization between Smo and the Cos2/Fu complex, which was prevented by phosphor-deficient mutations that lock Smo in the "close" conformation [31]. On the other hand, phosphor-mimetic mutations, which lock Smo in the "open" conformation, promotes colocalization between Smo and the Cos2/Fu complex [31]. Deletion study indicated that the Hh-induce conformational change in Smo C-tail exposes a Cos2-docking site(s) near the C-terminus of Smo, which facilitates the assembly of an active Smo-Cos2-Fu complex [31]. The conformational switch coupled with dimerization/oligomerization of Smo C-tails promotes dimerization/oligomerization of Fu, leading to its phosphorylation and activation [31]. Activated Fu regulates both Ci^R and Ci^A by controlling Ci-Sufu and Ci-Cos2-kinase complex formation [31].

Although the mammalian Fu homolog is not involved in Hh signaling, a recent study provided evidence that Kif7 forms a complex with an activated form of Smo depending on two ciliary proteins, Evc and Evc2, the products of human disease genes responsible for the Ellis-van Creveld syndrome [67]. It would be interesting to identify the Fu equivalent kinase(s) in the mammalian Hh pathway and determine whether the phosphorylation-induced conformational switch of mSmo activates this kinase through a mechanism similar to Fu activation.

2.4 Differential Phosphorylation of Smo Translates Hh Morphogen Gradient

A hallmark of Hh signaling is its ability to act over a long range with different levels of Hh ligand specifying distinct developmental outcomes [7, 10]. How different levels of the Hh signal are interpreted by receiving cells is still poorly understood.

However, recent studies suggest that graded Hh signals are translated into different levels of Smo activity via differential phosphorylation [9, 21, 25, 61]. In response to increasing levels of Hh ligand, both *Drosophila* and mammalian Smo proteins are phosphorylated at increasing levels [61, 68]. Increasing the number of phospho-mimetic mutations in Smo resulted in a progressive change of Smo cell surface/ciliary accumulation, C-tail conformation, and pathway activity [21, 25, 61]. In dSmo, phosphorylation of a proximal PKA/CK1 cluster can influence the phosphorylation of a distal cluster, implying that phosphorylation of dSmo C-tail may occur in a sequential manner [68]. Another study showed that the levels of Smo phosphorylation are dynamically regulated by phosphatases in *Drosophila*, with protein phosphatase 1 (PP1) dephosphorylating PKA-phosphorylated Smo to reduce signaling mediated by intermediate concentrations of Hh and protein phosphatase 2A (PP2A) specifically dephosphorylating PKA-primed, CK1-phosphorylated Smo to restrict signaling mediated by high levels of Hh [69].

In dSmo, the PKA/CK1 phosphorylation sites are located adjacent to the multiple Arg motifs that mediate negative regulation of Smo in the SAID domain (Fig. 2). Such an arrangement may allow more precise regulation of Smo activity because phosphorylation at a given PKA/CK1 cluster may only neutralize the negative influence of an adjacent Arg motifs, leading to an incremental change in Smo activity. Indeed, increasing the number of Arg to Ala (RA) mutation has a dosage effect on Smo activity similar to increasing the number of phospho-mimetic mutation, both of which progressively increased the levels of Smo activity [21, 25]. Hence, by employing multiple Arg clusters as inhibitory elements and modulating differential phosphorylation status of their adjacent PKA/CK1 sites, Smo can act as a rheostat to translate graded Hh signals into distinct responses [25].

2.5 *Phosphorylation-Regulated Ubiquitination Controls Smo Trafficking and Cell Surface Expression*

In *Drosophila*, Hh-induced phosphorylation promotes Smo cell surface accumulation but the underlying mechanism has remained a mystery until recently. Two studies revealed that dSmo undergoes ubiquitination [70, 71], another type of PTM that often modulates protein trafficking and stability. Ubiquitination is a process of attaching ubiquitin (Ub) molecules to a substrate protein, which involves activation by an Ub-activation enzyme (E1) and subsequent transfer by an Ub-conjugating enzyme (E2) in conjunction with an Ub ligase (E3) that recognizes and determines the specificity of the target protein [72]. While mono-ubiquitination (the addition of one ubiquitin molecule to one substrate protein residue) or multi-ubiquitination (the addition of one ubiquitin molecule to multiple residues of a substrate) mainly affects membrane trafficking and endocytosis of the target proteins [73, 74], poly-ubiquitination (the formation of a ubiquitin chain on a single lysine residue of the substrate protein) executes various effects including targeting substrate for proteasome- or lysosome-mediated degradation, regulating the subcellular localization of substrate, activating or inactivating proteins by modulating protein–protein interactions [72]. Similar to phosphorylation, ubiquitination is a reversible process that can be dynamically regulated by deubiquitinating enzymes (DUBs) [75].

In *Drosophila* wing discs, Smo cell surface level is low in anterior (A) compartment cells away from the A/P boundary but is elevated in response to Hh in A-compartment cells near the A/P boundary or in posterior (P) compartment cells. Inactivation of the ubiquitin-activating enzyme Uba1 led to ectopic Smo cell surface accumulation in A-compartment cells whereas loss of the deubiquitinating enzyme UBPY/USP8 attenuated Smo cell surface expression in P-compartment cells, suggesting that ubiquitination suppresses the Hh-induced Smo cell surface accumulation [70, 71]. In S2 cells, Smo is ubiquitinated at multiple Lys residues including those located in the SAID domain and both mono- and poly-ubiquitination have been implicated in the regulation of Smo trafficking [70, 71, 76]. Ubiquitination of Smo results in its endocytosis and degradation by both proteasome- and lysosome-dependent pathways. Hh inhibits Smo ubiquitination in a manner depending on PKA/CK1-mediated phosphorylation of Smo C-tail as well as the deubiquitinating enzyme UBPY/USP8, leading to Smo accumulation on the cell surface (Fig. 1) [70, 71].

mSmo is also ubiquitinated in NIH3T3 cells in a manner inhibited by Shh [70]; however, it is not clear whether ubiquitination regulates mSmo trafficking and/or ciliary accumulation. The E3 ubiquitin ligase(s) responsible for Smo ubiquitination has remained unknown in both *Drosophila* and mammals. Another important and unanswered question is how Hh and PKA/CK1-mediated phosphorylation inhibit Smo ubiquitination. Since the interaction between Smo and UBPY/USP8 is not significantly affected by either Hh stimulation or change in the Smo phosphorylation status in cultured cells [71], and UBPY/USP8 appears to catalyze Smo deubiquitination in both signal "off" and "on" states [70, 71], the Hh may inhibit Smo ubiquitination by regulating the recruitment of one or more E3 ubiquitin ligases or through other mechanisms. Hence, identifying the E3 ligase(s) involved in Smo ubiquitination may shed important light on the mechanism by which Smo trafficking is regulated.

2.6 Regulation of Smo Trafficking and Cell Surface Accumulation by Smo-Binding Proteins

In addition to ubiquitination, several Smo-binding proteins also regulate Smo trafficking. An earlier study revealed that the Cos2/Fu complex regulates Smo phosphorylation, thereby indirectly influencing Smo trafficking [77]. In *Drosophila* wing imaginal discs, loss-of-Fu function blocked the Hh-induced Smo accumulation in A compartment cell near the A/P boundary but this defect was suppressed by simultaneously removing Cos2 and Fu, suggesting that Fu promotes Smo cell surface accumulation by antagonizing a negative regulation imposed by Cos2 [77]. Cos2 recruits PP4 to dephosphorylate Smo, which needs to be reversed in order for Hh to promote hyper-phosphorylation and cell surface accumulation of Smo. Inhibition of Cos2/PP4 appears to be mediated by Fu phosphorylation of Cos2 at Ser572, leading to dissociation of Cos2/PP4 from a membrane proximal Cos2-binding domain in the Smo C-tail [77].

In the case of GPCRs, agonists induce receptor ubiquitination and recruit β-arrestin to downregulate GPCRs, leading to pathway desensitization [78, 79]. Interestingly, β-arrestin also regulates Smo trafficking in both *Drosophila* and mammals. The *Drosophila* non-visual β-arrestin Kurtz (Krz) formed a complex with Smo in S2 cells and its overexpression downregulated Smo cell surface accumulation in wing imaginal discs [71, 80]. However, loss of Krz in wing discs did not result in Smo accumulation, likely because ubiquitination can promote Smo endocytosis and degradation in the absence of Krz [71, 80]. Indeed, Krz RNAi in S2 cells increased the cell surface expression of an ubiquitination-deficient form of Smo [71]. Interestingly, Hh inhibits Smo/Krz association [71], suggesting that Hh promotes Smo cell surface accumulation, at least in part, by preventing Krz recruitment. By contrast, β-arrestin promotes ciliary accumulation of mSmo and its binding to mSmo is stimulated by Shh-induced phosphorylation of mSmo [61, 63].

In *Drosophila* wing discs, loss of Gprk2 leads to Smo cell surface accumulation in the absence of or in the presence of low levels of Hh ligand whereas overexpression of Gprk2 downregulates Smo cell surface accumulation [57, 58, 81]. Hence, Gprk2 plays a dual role in Hh signaling: it prevents aberrant Smo accumulation and pathway activation in the absence of Hh but is required for maximal Smo activation in the presence of high levels of Hh. Gprk2 regulates Smo level depending on its kinase activity [57], but the relevant substrate has remained elusive. In the case of GPCRs, GRK/Gprk phosphorylates the receptors after agonist stimulation, which recruits β-arrestin to downregulate GPCRs. However, Gprk2 and Krz appear to regulate Smo cell surface accumulation via independent mechanisms [80]. Therefore, the precise mechanisms by which Gprk2 and Krz regulate Smo trafficking and cell surface expression remain to be determined in the future.

3 Summary and Future Prospects

In this chapter, we review our current knowledge regarding the role of phosphorylation and ubiquitination as well as Smo-binding proteins in the regulation of Smo trafficking, subcellular localization, conformation, and signal transduction. It is noticeable that, although mammalian Hh pathway relies on the primary cilium and exhibits great divergence from the *Drosophila* pathway, recent studies have revealed a striking conservation in the mechanism underlying Smo activation

[60, 61, 82]. Many important questions remain to be addressed in the future. First, although several Smo kinases have been identified, they only account for half of the phosphorylation events identified in the Smo C-tail, kinases responsible for the remaining phosphorylation events and the full spectrum of phosphorylation-mediated regulatory mechanisms await to be explored. Second, the mechanism by which Ptc inhibits Smo phosphorylation has remained an enigma, and the physiological ligand(s) for Smo is still elusive. Third, the E3 ligase(s) that catalyzes Smo ubiquitination remains to be identified and how its activity is regulated by upstream signal awaits determination. Finally, it is possible that other types of PTM, in addition to phosphorylation and ubiquitination, may also contribute to the regulation of Smo. It is important to explore how different PTMs and Smo-binding proteins coordinate to precisely control Smo activity in response to graded Hh signals and whether these PTM-mediated regulatory events of Smo are conserved from flies to mammals.

Acknowledgments J.J. is supported by grants from National Institute of Health (GM061269 and GM067045), Cancer Prevention Research Institute of Texas (RP100561), and Welch Foundation (I-1603).

References

1. Chen MH, Li YJ, Kawakami T, Xu SM, Chuang PT (2004) Palmitoylation is required for the production of a soluble multimeric Hedgehog protein complex and long-range signaling in vertebrates. Genes Dev 18:641–659
2. Panakova D, Sprong H, Marois E, Thiele C, Eaton S (2005) Lipoprotein particles are required for Hedgehog and Wingless signalling. Nature 435:58–65
3. Zeng X, Goetz JA, Suber LM, Scott WJ Jr, Schreiner CM, Robbins DJ (2001) A freely diffusible form of Sonic hedgehog mediates long-range signalling. Nature 411:716–720
4. Huang S, Zhang Z, Zhang C, Lv X, Zheng X, Chen Z, Sun L, Wang H, Zhu Y, Zhang J, Yang S, Lu Y, Sun Q, Tao Y, Liu F, Zhao Y, Chen D (2013) Activation of Smurf E3 ligase promoted by smoothened regulates hedgehog signaling through targeting patched turnover. PLoS Biol 11:e1001721
5. Stone DM, Hynes M, Armanini M, Swanson TA, Gu Q, Johnson RL, Scott MP, Pennica D, Goddard A, Phillips H, Noll M, Hooper JE, de Sauvage F, Rosenthal A (1996) The tumour-suppressor gene *patched* encodes a candidate receptor for Sonic hedgehog. Nature 384:129–134
6. Jiang J (2006) Regulation of Hh/Gli signaling by dual ubiquitin pathways. Cell Cycle 5:2457–2463
7. Jiang J, Hui CC (2008) Hedgehog signaling in development and cancer. Dev Cell 15:801–812
8. Wilson CW, Chuang PT (2010) Mechanism and evolution of cytosolic Hedgehog signal transduction. Development 137:2079–2094
9. Chen Y, Jiang J (2013) Decoding the phosphorylation code in Hedgehog signal transduction. Cell Res 23:186–200
10. Ingham PW, McMahon AP (2001) Hedgehog signaling in animal development: paradigms and principles. Genes Dev 15:3059–3087
11. Nusslein-Volhard C, Wieschaus E (1980) Mutations affecting segment number and polarity in Drosophila. Nature 287:795–801

12. Alcedo J, Ayzenzon M, Von Ohlen T, Noll M, Hooper JE (1996) The *Drosophila smoothened* gene encodes a seven-pass membrane protein, a putative receptor for the Hedgehog signal. Cell 86:221–232
13. Van den Heuval M, Ingham PW (1996) Smoothened encodes a receptor-like serpentine protein required for hedgehog signalling. Nature 382:547–551
14. Rudin CM, Hann CL, Laterra J, Yauch RL, Callahan CA, Fu L, Holcomb T, Stinson J, Gould SE, Coleman B, LoRusso PM, Von Hoff DD, de Sauvage FJ, Low JA (2009) Treatment of medulloblastoma with hedgehog pathway inhibitor GDC-0449. N Engl J Med 361:1173–1178
15. Von Hoff DD, LoRusso PM, Rudin CM, Reddy JC, Yauch RL, Tibes R, Weiss GJ, Borad MJ, Hann CL, Brahmer JR, Mackey HM, Lum BL, Darbonne WC, Marsters JC Jr, de Sauvage FJ, Low JA (2009) Inhibition of the hedgehog pathway in advanced basal-cell carcinoma. N Engl J Med 361:1164–1172
16. Low JA, de Sauvage FJ (2010) Clinical experience with Hedgehog pathway inhibitors. J Clin Oncol 28:5321–5326
17. Kunstfeld R (2014) Smoothened inhibitors in the treatment of advanced basal cell carcinomas. Curr Opin Oncol 26:184–195
18. Bidet M, Joubert O, Lacombe B, Ciantar M, Nehme R, Mollat P, Bretillon L, Faure H, Bittman R, Ruat M, Mus-Veteau I (2011) The hedgehog receptor patched is involved in cholesterol transport. PLoS One 6:e23834
19. Taipale J, Cooper MK, Maiti T, Beachy PA (2002) Patched acts catalytically to suppress the activity of Smoothened. Nature 418:892–897
20. Denef N, Neubuser D, Perez L, Cohen SM (2000) Hedgehog induces opposite changes in turnover and subcellular localization of patched and smoothened. Cell 102:521–531
21. Jia J, Tong C, Wang B, Luo L, Jiang J (2004) Hedgehog signalling activity of smoothened requires phosphorylation by protein kinase a and casein kinase I. Nature 432:1045–1050
22. Zhu AJ, Zheng L, Suyama K, Scott MP (2003) Altered localization of Drosophila Smoothened protein activates Hedgehog signal transduction. Genes Dev 17:1240–1252
23. Corbit KC, Aanstad P, Singla V, Norman AR, Stainier DY, Reiter JF (2005) Vertebrate Smoothened functions at the primary cilium. Nature 437:1018–1021
24. Rohatgi R, Milenkovic L, Scott MP (2007) Patched1 regulates hedgehog signaling at the primary cilium. Science 317:372–376
25. Zhao Y, Tong C, Jiang J (2007) Hedgehog regulates smoothened activity by inducing a conformational switch. Nature 450:252–258
26. Jia J, Tong C, Jiang J (2003) Smoothened transduces Hedgehog signal by physically interacting with Costal2/Fused complex through its C-terminal tail. Genes Dev 17:2709–2720
27. Lum L, Zhang C, Oh S, Mann RK, von Kessler DP, Taipale J, Weis-Garcia F, Gong R, Wang B, Beachy PA (2003) Hedgehog signal transduction via Smoothened association with a cytoplasmic complex scaffolded by the atypical kinesin, Costal-2. Mol Cell 12:1261–1274
28. Ogden SK, Ascano M Jr, Stegman MA, Suber LM, Hooper JE, Robbins DJ (2003) Identification of a functional interaction between the transmembrane protein Smoothened and the kinesin-related protein Costal2. Curr Biol 13:1998–2003
29. Ruel L, Gallet A, Raisin S, Truchi A, Staccini-Lavenant L, Cervantes A, Therond PP (2007) Phosphorylation of the atypical kinesin Costal2 by the kinase fused induces the partial disassembly of the Smoothened-Fused-Costal2-Cubitus interruptus complex in Hedgehog signalling. Development 134:3677–3689
30. Zhang W, Zhao Y, Tong C, Wang G, Wang B, Jia J, Jiang J (2005) Hedgehog-regulated costal2-kinase complexes control phosphorylation and proteolytic processing of cubitus interruptus. Dev Cell 8:267–278
31. Shi Q, Li S, Jia J, Jiang J (2011) The Hedgehog-induced Smoothened conformational switch assembles a signaling complex that activates fused by promoting its dimerization and phosphorylation. Development 138:4219–4231
32. Ohlmeyer JT, Kalderon D (1998) Hedgehog stimulates maturation of Cubitus interruptus into a labile transcriptional activator. Nature 396:749–753

33. Cheung HO, Zhang X, Ribeiro A, Mo R, Makino S, Puviindran V, Law KK, Briscoe J, Hui CC (2009) The kinesin protein Kif7 is a critical regulator of Gli transcription factors in mammalian hedgehog signaling. Sci Signal 2:ra29
34. Liem KF Jr, He M, Ocbina PJ, Anderson KV (2009) Mouse Kif7/Costal2 is a cilia-associated protein that regulates Sonic hedgehog signaling. Proc Natl Acad Sci U S A 106:13377–13382
35. Endoh-Yamagami S, Evangelista M, Wilson D, Wen X, Theunissen JW, Phamluong K, Davis M, Scales SJ, Solloway MJ, de Sauvage FJ, Peterson AS (2009) The mammalian Cos2 homolog Kif7 plays an essential role in modulating Hh signal transduction during development. Curr Biol 19:1320–1326
36. Law KK, Makino S, Mo R, Zhang X, Puviindran V, Hui CC (2012) Antagonistic and cooperative actions of Kif7 and Sufu define graded intracellular Gli activities in Hedgehog signaling. PLoS One 7:e50193
37. Maurya AK, Ben J, Zhao Z, Lee RT, Niah W, Ng AS, Iyu A, Yu W, Elworthy S, van Eeden FJ, Ingham PW (2013) Positive and negative regulation of Gli activity by Kif7 in the zebrafish embryo. PLoS Genet 9:e1003955
38. Chen MH, Gao N, Kawakami T, Chuang PT (2005) Mice deficient in the fused homolog do not exhibit phenotypes indicative of perturbed hedgehog signaling during embryonic development. Mol Cell Biol 25:7042–7053
39. Merchant M, Evangelista M, Luoh SM, Frantz GD, Chalasani S, Carano RA, van Hoy M, Ramirez J, Ogasawara AK, McFarland LM, Filvaroff EH, French DM, de Sauvage FJ (2005) Loss of the serine/threonine kinase fused results in postnatal growth defects and lethality due to progressive hydrocephalus. Mol Cell Biol 25:7054–7068
40. Maloverjan A, Piirsoo M, Michelson P, Kogerman P, Osterlund T (2010) Identification of a novel serine/threonine kinase ULK3 as a positive regulator of Hedgehog pathway. Exp Cell Res 316:627–637
41. Maloverjan A, Piirsoo M, Kasak L, Peil L, Osterlund T, Kogerman P (2010) Dual function of UNC-51-like kinase 3 (Ulk3) in the Sonic hedgehog signaling pathway. J Biol Chem 285:30079–30090
42. Ogden SK, Fei DL, Schilling NS, Ahmed YF, Hwa J, Robbins DJ (2008) G protein Galpha (i) functions immediately downstream of Smoothened in Hedgehog signalling. Nature 456:967–970
43. Low WC, Wang C, Pan Y, Huang XY, Chen JK, Wang B (2008) The decoupling of Smoothened from Galphai proteins has little effect on Gli3 protein processing and Hedgehog-regulated chick neural tube patterning. Dev Biol 321:188–196
44. Cozzone AJ (1988) Protein phosphorylation in prokaryotes. Annu Rev Microbiol 42:97–125
45. Stock JB, Ninfa AJ, Stock AM (1989) Protein phosphorylation and regulation of adaptive responses in bacteria. Microbiol Rev 53:450–490
46. Chang C, Stewart RC (1998) The two-component system. Regulation of diverse signaling pathways in prokaryotes and eukaryotes. Plant Physiol 117:723–731
47. Barford D, Das AK, Egloff MP (1998) The structure and mechanism of protein phosphatases: insights into catalysis and regulation. Annu Rev Biophys Biomol Struct 27:133–164
48. Jiang J, Struhl G (1995) Protein kinase A and Hedgehog signalling in *Drosophila* limb development. Cell 80:563–572
49. Li W, Ohlmeyer JT, Lane ME, Kalderon D (1995) Function of protein kinase A in hedghehog signal transduction and Drosophila imaginal disc development. Cell 80:553–562
50. Wang G, Wang B, Jiang J (1999) Protein kinase A antagonizes Hedgehog signaling by regulating both the activator and repressor forms of Cubitus interruptus. Genes Dev 13:2828–2837
51. Jia J, Zhang L, Zhang Q, Tong C, Wang B, Hou F, Amanai K, Jiang J (2005) Phosphorylation by double-time/CKIepsilon and CKIalpha targets cubitus interruptus for Slimb/beta-TRCP-mediated proteolytic processing. Dev Cell 9:819–830

52. Price MA, Kalderon D (2002) Proteolysis of the Hedgehog signaling effector Cubitus interruptus requires phosphorylation by Glycogen Synthase Kinase 3 and Casein Kinase 1. Cell 108:823–835
53. Ohlmeyer JT, Kalderon D (1997) Dual pathways for induction of wingless expression by protein kinase A and Hedgehog in Drosophila embryos. Genes Dev 11:2250–2258
54. Apionishev S, Katanayeva NM, Marks SA, Kalderon D, Tomlinson A (2005) Drosophila Smoothened phosphorylation sites essential for Hedgehog signal transduction. Nat Cell Biol 7:86–92
55. Zhang C, Williams EH, Guo Y, Lum L, Beachy PA (2004) Extensive phosphorylation of Smoothened in Hedgehog pathway activation. Proc Natl Acad Sci U S A 101:17900–17907
56. Jia H, Liu Y, Xia R, Tong C, Yue T, Jiang J, Jia J (2010) Casein kinase 2 promotes Hedgehog signaling by regulating both smoothened and Cubitus interruptus. J Biol Chem 285:37218–37226
57. Chen Y, Li S, Tong C, Zhao Y, Wang B, Liu Y, Jia J, Jiang J (2010) G protein-coupled receptor kinase 2 promotes high-level Hedgehog signaling by regulating the active state of Smo through kinase-dependent and kinase-independent mechanisms in Drosophila. Genes Dev 24:2054–2067
58. Cheng S, Maier D, Neubueser D, Hipfner DR (2010) Regulation of smoothened by Drosophila G-protein-coupled receptor kinases. Dev Biol 337:99–109
59. Chen W, Ren XR, Nelson CD, Barak LS, Chen JK, Beachy PA, de Sauvage F, Lefkowitz RJ (2004) Activity-dependent internalization of smoothened mediated by beta-arrestin 2 and GRK2. Science 306:2257–2260
60. Meloni AR, Fralish GB, Kelly P, Salahpour A, Chen JK, Wechsler-Reya RJ, Lefkowitz RJ, Caron MG (2006) Smoothened signal transduction is promoted by G protein-coupled receptor kinase 2. Mol Cell Biol 26:7550–7560
61. Chen Y, Sasai N, Ma G, Yue T, Jia J, Briscoe J, Jiang J (2011) Sonic Hedgehog dependent phosphorylation by CK1a and GRK2 is required for ciliary accumulation and activation of Smoothened. PLoS Biol 9:e1001083
62. Evangelista M, Lim TY, Lee J, Parker L, Ashique A, Peterson AS, Ye W, Davis DP, de Sauvage FJ (2008) Kinome siRNA screen identifies regulators of ciliogenesis and hedgehog signal transduction. Sci Signal 1:ra7
63. Kovacs JJ, Whalen EJ, Liu R, Xiao K, Kim J, Chen M, Wang J, Chen W, Lefkowitz RJ (2008) Beta-arrestin-mediated localization of smoothened to the primary cilium. Science 320:1777–1781
64. Rohatgi R, Milenkovic L, Corcoran RB, Scott MP (2009) Hedgehog signal transduction by Smoothened: pharmacologic evidence for a 2-step activation process. Proc Natl Acad Sci U S A 106:3196–3201
65. Wang Y, Zhou Z, Walsh CT, McMahon AP (2009) Selective translocation of intracellular Smoothened to the primary cilium in response to Hedgehog pathway modulation. Proc Natl Acad Sci U S A 106:2623–2628
66. Wilson CW, Chen MH, Chuang PT (2009) Smoothened adopts multiple active and inactive conformations capable of trafficking to the primary cilium. PLoS One 4:e5182
67. Yang C, Chen W, Chen Y, Jiang J (2012) Smoothened transduces Hedgehog signal by forming a complex with Evc/Evc2. Cell Res 22:1593–1604
68. Fan J, Liu Y, Jia J (2012) Hh-induced Smoothened conformational switch is mediated by differential phosphorylation at its C-terminal tail in a dose- and position-dependent manner. Dev Biol 366:172–184
69. Su Y, Ospina JK, Zhang J, Michelson AP, Schoen AM, Zhu AJ (2011) Sequential phosphorylation of smoothened transduces graded hedgehog signaling. Sci Signal 4:ra43
70. Xia R, Jia H, Fan J, Liu Y, Jia J (2012) USP8 promotes smoothened signaling by preventing its ubiquitination and changing its subcellular localization. PLoS Biol 10:e1001238
71. Li S, Chen Y, Shi Q, Yue T, Wang B, Jiang J (2012) Hedgehog-regulated ubiquitination controls smoothened trafficking and cell surface expression in Drosophila. PLoS Biol 10: e1001239

72. Komander D, Rape M (2012) The ubiquitin code. Annu Rev Biochem 81:203–229
73. Miranda M, Sorkin A (2007) Regulation of receptors and transporters by ubiquitination: new insights into surprisingly similar mechanisms. Mol Interv 7:157–167
74. Hicke L (2001) Protein regulation by monoubiquitin. Nat Rev Mol Cell Biol 2:195–201
75. Reyes-Turcu FE, Ventii KH, Wilkinson KD (2009) Regulation and cellular roles of ubiquitin-specific deubiquitinating enzymes. Annu Rev Biochem 78:363–397
76. Yang X, Mao F, Lv X, Zhang Z, Fu L, Lu Y, Wu W, Zhou Z, Zhang L, Zhao Y (2013) Drosophila Vps36 regulates Smo trafficking in Hedgehog signaling. J Cell Sci 126:4230–4238
77. Liu Y, Cao X, Jiang J, Jia J (2007) Fused-Costal2 protein complex regulates Hedgehog-induced Smo phosphorylation and cell-surface accumulation. Genes Dev 21:1949–1963
78. Shenoy SK, McDonald PH, Kohout TA, Lefkowitz RJ (2001) Regulation of receptor fate by ubiquitination of activated beta 2-adrenergic receptor and beta-arrestin. Science 294:1307–1313
79. Shenoy SK, Xiao K, Venkataramanan V, Snyder PM, Freedman NJ, Weissman AM (2008) Nedd4 mediates agonist-dependent ubiquitination, lysosomal targeting, and degradation of the beta2-adrenergic receptor. J Biol Chem 283:22166–22176
80. Molnar C, Ruiz-Gomez A, Martin M, Rojo-Berciano S, Mayor F, de Celis JF (2011) Role of the Drosophila non-visual ss-arrestin kurtz in hedgehog signalling. PLoS Genet 7:e1001335
81. Molnar C, Holguin H, Mayor F Jr, Ruiz-Gomez A, de Celis JF (2007) The G protein-coupled receptor regulatory kinase GPRK2 participates in Hedgehog signaling in Drosophila. Proc Natl Acad Sci U S A 104:7963–7968
82. Zhao Y, Tong C, Jiang J (2007) Transducing the Hedgehog signal across the plasma membrane. Fly (Austin) 1:333–336

Top Med Chem (2015) 16: 127–146
DOI: 10.1007/7355_2014_64

Published online: 20 July 2014

Three-Dimensional Structure of the Smoothened Receptor: Implications for Drug Discovery

Didier Rognan and Isabelle Mus-Veteau

Abstract The recently described high resolution three-dimensional structures of the transmembrane and the extracellular domains of the human Smoothened (Smo) receptor higlight both conserved and unique structural features of this G protein-coupled receptor. It enables a better understanding of very subtle molecular mechanisms regulating Smo function and demonstrates the very plastic nature of this receptor which is able to accommodate a diverse array of small molecular weight ligands through several binding sites. This structural information should pave the way for designing small molecular weight modulators of Smo function targeting different binding sites and insensitive to clinically observed receptor mutations.

Keywords Binding mode, Cystein-rich domain, Drug design, Transmembrane domain, X-ray structure

Contents

D. Rognan (✉)
Laboratoire d'Innovation Thérapeutique, UMR 7200 CNRS-Université de Strasbourg, F-67400 Illkirch, France
e-mail: rognan@unistra.fr

I. Mus-Veteau
Institut de Pharmacologie Moléculaire et Cellulaire, UMR 7275 CNRS-Université de Nice-Sophia Antipolis, 67400 Illkirch, France

Abbreviations

BCC	Basal cell carcinoma
Bud	Budesonide
drSmo	Drosophila Smoothened
ECD	Extracellular domain
ECD	Cystein-rich domain
ECL	Extracellular loop
GPCR	G protein-coupled receptor
hSmo	Human Smoothened
Hh	Hedgehog
ICL	Intracellular loop
NMR	Nuclear magnetic resonance
Smo	Smoothened
TM	Transmembrane
zSmo	Zebrafish Smoothened

1 Introduction

The Hedgehog (Hh) signaling pathway plays a crucial role in the regulation of embryonic development [1]. Since its activation, due to point mutations of genes coding for components involved in its signal transduction, is linked to tumorigenesis [2], the Hh pathway has received considerable attention to develop novel antitumoral drugs. Among the many targets involved in this peculiar pathway, Smoothened (Smo), a seven transmembrane (TM) receptor, has been the target of many drug discovery programs because of its remote sequence similarity to the superfamily of G protein-coupled receptors (GPCRs), the largest subclass of macromolecules currently targeted by low molecular weight drugs [3]. GPCRs being highly druggable targets, Smo has focussed the attention on many drug hunters and pharmaceutical companies, leading to the recent launch in 2012 of vismodegib, the first Smo antagonist for the treatment of locally advanced and metastatic basal cell carcinoma (BCC) [4]. Several Smo antagonists are currently under clinical investigation for treating diverse type of cancers [5, 6]. The initial enthusiasm raised by the efficacy of vismodegib on BCCs has been tempered by disappointing clinical results due to Smo mutation inducing early tumor resistance [7]. To improve the clinical benefit of Smo antagonists, there is a need to better understand the Hh pathway and the fine molecular mechanisms leading to either Smo activation or inhibition.

Table 1 Available 3D structures of Smo in the Protein DataBank

PDB id	Species	Domain	Residues	Ligand	Reference
4C79	Zebrafish	ECD	41–158	–	[9]
4C7A	Zebrafish	ECD	41–158	–	[9]
2MAH	Drosophila	ECD	85–202	Budesonide	[8]
4JKV	Human	TM	190–493	LY2940680	[7]
4N4W	Human	TM	190–555	SANT-1	[11]
4O9R	Human	TM	190–555	Cyclopamine	[12]

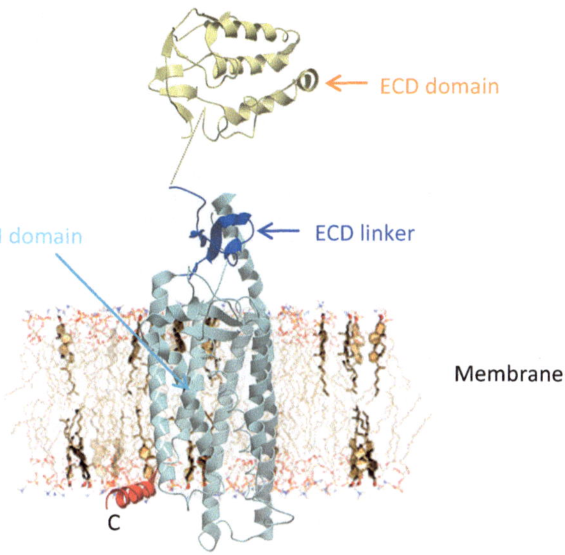

Fig. 1 Structure of human Smo. The extracellular ECD (*yellow*) is linked by a short linker (*blue*) to the TM domain (*cyan*) passing the membrane. The cytoplasmic domain begins with a short alpha helix (*red*) but the rest of its structure is still unknown. The still unsolved full structure is just presented for display purpose by connecting (*yellow dots*) the individual X-ray structure of the ECD [10] and TM [8] domains

The recent high resolution X-ray/NMR structures of the transmembrane (TM) [8] and the extracellular (EC) [9, 10] domains of Smo (Table 1) offer us the opportunity to review the structural peculiarities of this receptor, its subtle molecular regulation, and the future perspectives in designing a second generation of safer Smo modulators in various therapeutical indications.

The full aminoacid sequence of the human Smo (hSmo) receptor entails 787 amino acids (Uniprot Identifier: Q99385) organized in three main domains (Fig. 1): (1) the ECD (residues 1–191), (2) a seven heptahelical TM domain (residues 224–534), (3) a cytoplasmic domain (residues 535–787; Fig. 1). Although its canonical coupling to a G Protein has long been a matter of debate [13], the receptor was classified as a probable G protein-coupled receptor in the Frizzled family [14] to which hSmo shows the highest sequence identity.

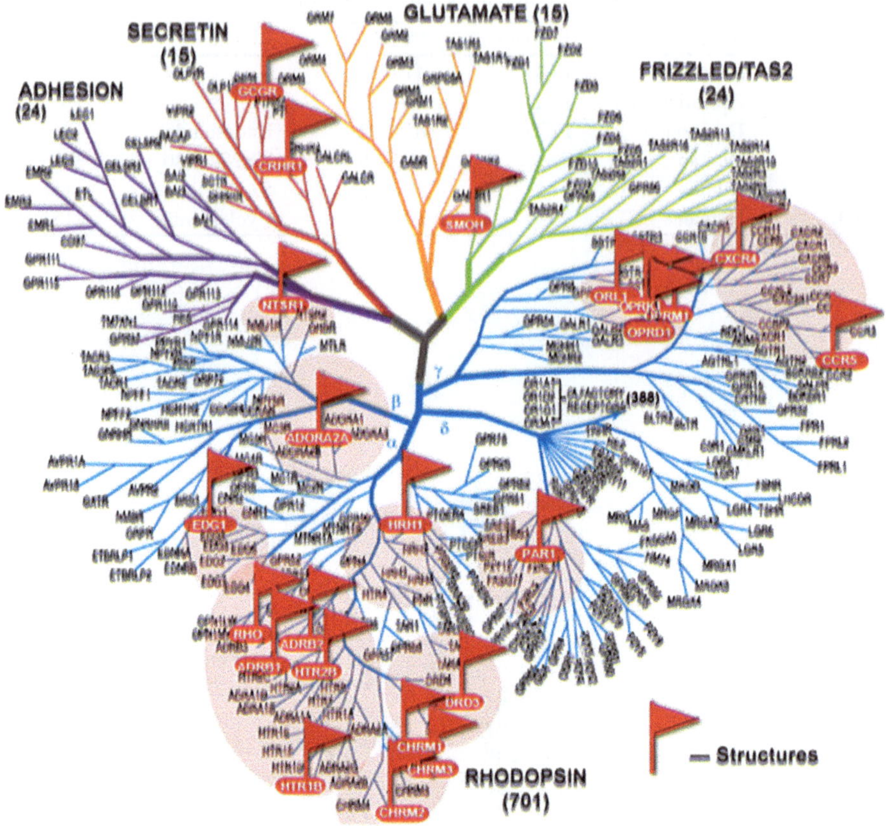

Fig. 2 Phylogenetic tree and structural coverage of human GPCRs (taken from the GPCR Network home page, [15])

Smo is one of the 23 GPCRs for which a high-resolution X-ray structure of the TM is now available (Fig. 2)

2 The Extracellular Domain

2.1 3D Structure

The N-terminal ECD of the Smo receptor is composed of a cysteine-rich domain (CRD) and an unusually long ECD linker domain. The X-ray structure of hSmo [8] does not show the structure of the full ECD since the 190 N-terminal aminoacids were replaced by a thermostabilized apocytochrome b562RIL (BRIL). However, the structure of the ECD linker domain is well resolved and shows that, when the

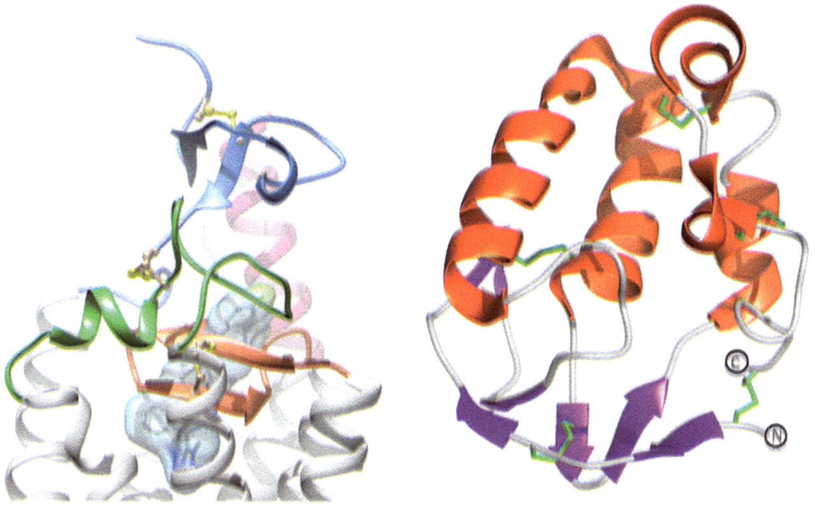

Fig. 3 (**a**) Structure of the ECD linker (*blue*) connected to the three extracellular loops (ECL1, *green*; ECL2, *orange*; ECL3, *pink*). Four conserved disulphide bonds (*ball* and *sticks*) maintain the overall stability of the ensemble. The bound-antagonist (LY2940680, transparent surface) sits mainly within the 7 TM bundle (*gray helices*), (**b**) Structure of the ECD domain of zebrafish Smo. α-helices, β-strands, and coils are displayed in *red*, *magenta*, and *gray*, respectively. The five disulphide bridges are represented as *green sticks*. N- and C-terminal residues are *circled*

small-antagonist LY2940680 is bound to Smo, the ECD linker domain is connected to the peculiarly long extracellular loops (ECLs) through four disulphide bonds and a hydrogen-bonding network (Fig. 3a). The cysteines that form disulphide bonds maintaining the connection between the ECD linker domain and the ECLs in the Smo receptor are highly preserved in the Frizzled (Fz) receptors.

Nachtergaele et al. [10] first crystallized and solved the structure of a portion comprised between residues 41 and 158 of the ECD of zebrafish Smo (zSmo) receptor which shows similarity to the Fz protein family ECD [16]. The zSmo ECD adopts a globular fold composed of four α-helices and a short two-stranded β-sheet stabilized by five disulphide bridges (Fig. 3b). This tertiary fold of Smo ECD is also observed in the drosophila Smo (drSmo) ECD whose structure was determined in solution using NMR spectroscopy [9]. Interestingly, the ECDs of Frizzled 8 (Fz8), secreted Frizzled-related protein 3 (sFRP3), muscle-specific kinase (MuSK), Niemann-Pick C1 protein (NPC1), riboflavin-binding protein (RFBP), and folate receptor α (FRα) adopt a similar helical bundle formed by helices α1, α2, and α3, and the four conserved disulphide bonds that stabilize the fold and the relative orientations of the helices (Fig. 4) [10].

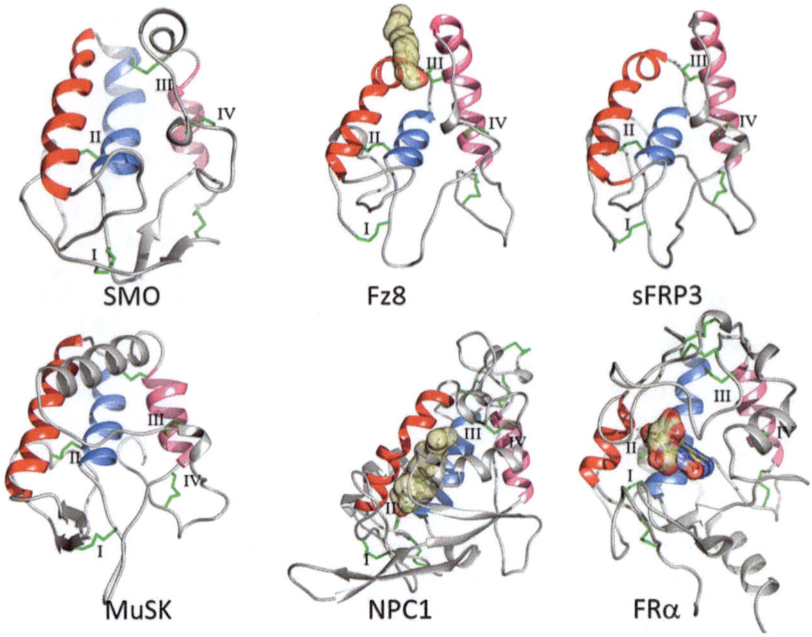

Fig. 4 Conserved structural pattern across ECDs of six representative targets: Smoothened (Smo, PDB ID 4C7A), Frizzled 8 (Fz8, PDBID 4F0A), secreted Frizzled related protein 3 (sFRP3, PDB ID 1IJX), muscle specific kinase (MuSK, PDB ID 3HKL), Niemann-Pick C1 protein (NPC1, PDB ID 3GKI), and folate receptor α (Frα, PDB ID 4LRH). Three α-helices (α-1, *red*, α-2, *blue*; α-3, *pink*) delimit the crevice (tan solid surface) to which the ligand (*sticks*) binds. A set of four conserved disulphide bridges (I–IV) rigidify the overall structure of the domain

2.2 Role on Smo Dimerization, Conformation, and Localization

The ECD of Frizzled has been shown to play an essential role in Wnt ligand binding and receptor dimerization [16, 17]. Although multi-angle light scattering analysis revealed that purified zSmo ECD behaved as a monomer in solution [10], the asymmetric unit of the zSmo crystal shows the presence of two zSmo molecules [8], suggesting that Smo ECD is involved in Smo dimerization.

The ECDs of the Frizzled GPCRs possess modest homology to that of Smo; nevertheless, the cysteines in these domains are highly conserved. To study the structural and biological significance of the disulphide bonds formed by the eight cysteines of the drSmo ECD, Rana et al. [9] replaced each cysteine by alanine, and observed that two of the four disulphide bridges play critical roles in maintaining the integrity of the ECD structure and the Hh signal transduction. DrSmo also possesses conserved cysteines in its ECD linker domain. The mutation of the cysteines of drSmo ECD that are homologous to the cysteines shown to form disulphide bonds between ECD and ECLs in hSmo structure failed to rescue Hh-induced reporter gene induction following endogenous Smo knockdown [9]. Thus, the disulphide bridges formed by the highly conserved ECD cysteines are essential for Smo conformation and downstream signaling.

Several recent studies have been carried out on ECD deletion mutants of Smo (SmoΔECD) both in drSmo [9, 18] and in vertebrate Smo [10, 19–21]. In drosophila, SmoΔECD mutants fail to dimerize suggesting that the ECD may govern Smo dimerization. DrSmoΔECD is inactive in contrast to vertebrate SmoΔECD. Vertebrate SmoΔECD shows a higher level of basal accumulation in cilia and a higher level of basal activity in Hh reporter assays suggesting that Smo ECD represses Hh signaling in the basal state. This basal activity is inhibited by Hh receptor Patched co-expression, suggesting that Smo regulation by Patched does not intervene at the ECD level. SmoΔECD remains responsive to SAG, a small agonist molecule that binds to the Smo TM, suggesting that the ECD is not absolutely required for signaling in response to synthetic Smo agonists. However, SmoΔECD retains only a low level of Hh responsiveness indicating that the ECD is required for Smo to become fully activated in response to Hh. The SmoΔECD localizes to the cilia but is unable to activate high-level Hh signaling. Data indicate that the ECD both regulates Smo ciliary localization and is essential for high-level Hh signaling.

2.3 Role on Ligand Binding

Classical GPCR functioning involves binding of ligands to the extracellular region of the receptor which induces conformational change in the core of the TM to modulate downstream signaling. A common feature of the Fz-like ECD family members is their ability to bind small hydrophobic molecules in a pocket formed by the core helices α1, α2, and α3 (Fig. 4). Fz8 ECD, which is the closest structural homologue of Smo ECD, binds to the endogenous ligand Wnt at two sites [16, 17]. The palmitoleyl moiety of Wnt binds in a shallow groove on the Fz8 ECD termed site 1, whereas the second Wnt-binding site on Fz8 ECD is located on the opposite end and involves protein–protein interactions. It has been shown that Smo-mediated signal transduction is highly sensitive to genetic or pharmacological depletion of endogenous cellular sterols [22–24] and that Smo can be activated by noncellular hydroxylated cholesterol derivatives such as 20(S)-hydroxycholesterol (20(S)-OHC) [10, 23, 25, 26]. Hydroxysterols and glucocorticoids were reported to modulate Smo signaling activity by binding to an Smo domain distinct from the orthosteric site present in the TM shown to bind the antagonist cyclopamine [26, 27]. Recent reports indicate that 20(S)-OHC binds to the Smo ECD of vertebrate Smo [20, 21] and the glucocorticoid budesonide (Bud) binds to the Smo ECDs from both drosophila and human [9], involving a binding cleft structurally analogous to that recognizing the Wnt palmitate adduct in Fz8.

Four papers published in 2013 demonstrated that the binding site for oxysterols and/or glucocorticoids maps to the extracellular ECD of both drosophila and vertebrate Smo and is completely separable from the site bound by other small-molecule Smo modulators located within the TM [9, 10, 20, 21]. Vertebrate SmoΔECD mutants no longer respond to oxysterol and cannot be fully activated by the Hh ligand [10, 20, 21]. In vertebrate Smo, responsiveness to 20(S)-OHC was not affected by the TM cyclopamine pocket mutations previously shown to block the effects of cyclopamine-competitive Smo agonists and antagonists suggesting that the TM domain might not contain the binding site for hydroxyl-sterols. Accordingly, Smo depleted of its ECD shows no responsiveness to induction by 20(S)-OHC and reduced responsiveness to induction by Hh [21].

Moreover, the ECD is not required for Smo activation by TM-binding synthetic agonists such as SAG, suggesting that the ECD has only a modulating role in response to Hh [20]. DrSmo does not bind oxysterols and is not inhibited by sterol depletion [9, 20]. However, Rana et al. [9] recently demonstrated that the glucocorticoid Bud binds to drSmo and suggested that Bud could act as a weak synthetic antagonist of an as yet unidentified endogenous ligand for the drSmo ECD, displacing it from its binding pocket.

2.4 Hypothesis for a Physiological Mechanism of Smo Regulation by Its ECD

A key question in the study of Hh signal transduction concerns the physiological mechanisms that regulate Smo activity in Hh-responsive cells. One leading model proposes that Smo is regulated by an endogenous lipidic modulator whose availability is controlled by Patched [28, 29] and by binding to the ECD or the TM domain (or both). Although 20(S)-OHC is capable of activating Smo when supplied exogenously to Hh-responsive cells, no sterol hydroxylase has ever been reported to produce 20(S)-OHC in cells and endogenous 20(S)-OHC could never be detected in cells [21, 30]. Thus, the physiological relevance of this lipid in the context of Hh pathway stimulation is not clear. The existence of common Smo and Fz structural elements implies that some aspects of their activation mechanisms may be fundamentally related. In vivo, active Hh is modified by a palmitate in its N-terminal extremity and cholesterol in its C-terminal end. Does the cholesterol or the palmitate moiety of Hh interact with Smo ECD in a comparable manner as Wnt with its receptor Fz8, and induce the conformational state necessary for full Hh signaling? Bidet et al. recently demonstrated that Patched effluxes cholesterol from Hh-responsive cells decreasing their intracellular cholesterol concentration [24]. Hh induces an increase of the intracellular cholesterol concentration in Hh-responsive cells triggering Smo accumulation to the plasma membrane. The cholesterol transport activity of Patched could be responsible for the inhibition of the accumulation of Smo in the cilia and thus for the Hh signaling modulation.

These data are in good agreements with studies reporting that Smo mediated signal transduction is highly sensitive to genetic or pharmacological depletion of endogenous cellular sterols [22, 23].

3 The Transmembrane Domain

3.1 Overall Topology

Thanks to significant advances in the engineering, purification, and crystallization of stable and functionally active membrane proteins [31], this target family of utmost importance in pharmaceutical research has seen a dramatic increase in the number of representative X-ray structures from a single structure (bovine rhodopsin, [32]) in June 2000 to 105 PDB structures (23 different receptors, [33, 34] in February 2014. Since most subfamilies are now covered (Fig. 2), both common and peculiar structural features of Smo can be discussed in detail.

Despite a very low sequence identity (<10 %) to classical class A GPCRs, Smo exhibits seven TM helices connected by three extracellular loops (ECL) and three intracellular loops (ICL). Another conserved feature, totally unexpected before the resolution of the first X-ray structure of bovine rhodopsin, is the presence of a eighth helix (H8) at the beginning of the intracellular C-terminal tail, which serves as anchoring point to the membrane. Various roles have been reported for this helix including ligand binding [35], G protein coupling [36], receptor phosphorylation [37], β-arrestin recruitment [38], and receptor internalization [39].

However, Smo also presents noticeable peculiarities. First, an intricate network of disulphide bonds and non-bonded contacts holds the ECD to the three ECLs, thereby limiting the space necessary for any ligand to enter the TM cavity. Conversely, the TM cavity is wide open for class A and class B peptide-binding receptors. Second, the specific microdomains [40] (e.g., DRY in TM3, NPxxY in TM7) responsible for the sequential activation of class A receptors are all absent in Smo. The distribution of proline and glycine residues of the 7-TM bundle drastically differs strikingly (Fig. 5) suggesting class-specific activation mechanism for class F GPCRs.

Smo shows a high degree of similarity to class F GPCRs at transmembrane helices with 45 fully conserved residues [8]. Interestingly, the disulphide bonds participating to the structural stability of the ECD linker and of the three ECLs are conserved among Frizzled receptors. Close to the intracellular membrane, a cluster of tryptophan residues (W331, W339, W365, W535) are also strictly conserved, the mutation of one of them (W535, TM7) leading to a constitutively active Smo receptor [41]. Last, the typical TxxxW motif of helix 8, known to be critical for the interaction of Frizzled receptors with Disheveled and the activation of the Wnt/β-catenin signaling pathway [42], is also conserved in Smo, therefore enabling to properly anchor the helix 8 parallel to the membrane, and interact with

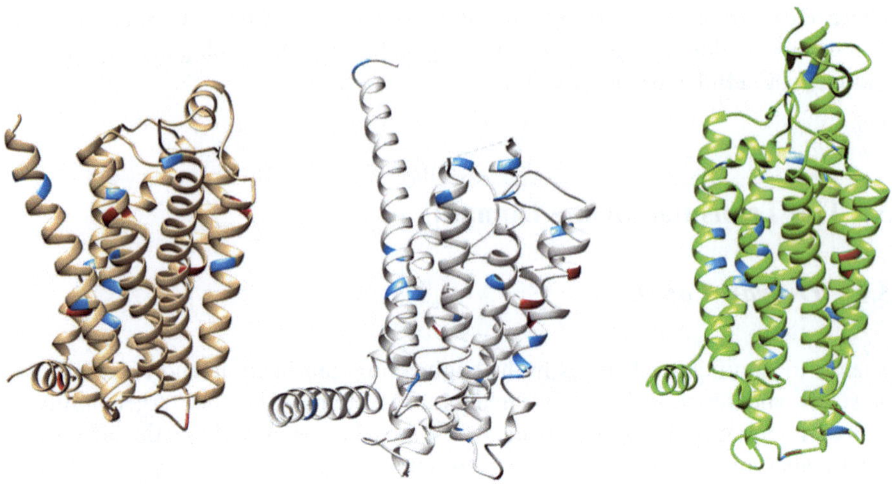

Fig. 5 3D structures of a class A (human beta2 adrenergic receptor, PDB ID 2rh1, tan), a class B (human corticotropin releasing factor type 1, PDB ID 4K5Y, *white*), and a class F (human Smoothened, PDB ID 4JKV, *green*). For each receptor, proline and glycine residues are displayed in *red* and *blue*, respectively

intracellular partners. The C-terminal tail, whose structure is unsolved in all X-ray structures of Smo is notably known to mediate Hh-induced phosphorylation by GRK2 which promotes a conformational switch of the receptor along with the dimerization of its C-terminal tail [18, 43].

3.2 Ligand-Binding Pocket

The precise location of the Smo TM cavity has long been obscure. Early truncation experiments of the full length-Smo clearly evidenced the TM domain as binding site for the first Smo antagonist cyclopamine [44]. Early clinical trials and experiments conducted on animal models with Smo antagonists showed that medulloblastoma readily escape chemotherapy by specific point mutations mostly occurring at the TM domain [7, 45]. The recent X-ray structure of the TM now provides a clear picture of the situation. As most GPCRs, Smo presents a deep hydrophobic cavity between the 7-TM bundle and the three ECLs, to which small molecular weight ligands bind.

Conversely to biogenic amine receptor ligands (adrenergic, dopamine, histamine, muscarinic receptors) Smo antagonists bind along an axis parallel to that of the 7-TM bundle. It is however striking to notice that the position of ligand binding to Smo recalls more that of some class A receptor ligands (adenosine A2a receptor, protease-activated receptor 1) than that of a class B antagonist

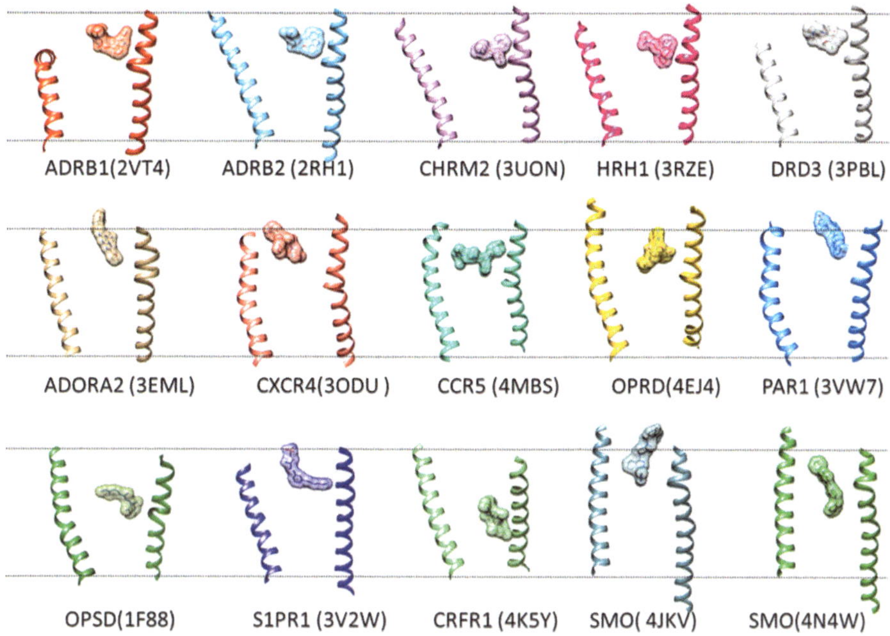

Fig. 6 Positioning of low molecular weight antagonists/inverse agonists in the 7-TM bundle of crystallized GPCRs. Only TM1 (*left hand side*) and TM5 (*right hand side*) are represented to delimit the receptor boundaries within the membrane (*dotted lines*). Examples are given for the following complexes: beta1 adrenergic (ADRB1)-cyanopindolol, beta2 adrenergic (ADRB2)-carazolol, muscarinic M2 (CHRM2)-QNB, histamine H1 (HRH1)-doxepin, dopamine D3 (D3DR)-eticlopride, Adenosine A2A (ADORA2)-ZM341285, chemokine CXCR4 (CXCR4)-IT1t, chemokine CCR5 (CCR5)-maraviroc, delta opioid (OPRD)-naltrindole, protease-activated receptor 1 (PAR1)-vorapaxar, bovine rhodopsin (OPSD)-11-*cis*-retinal, sphingosine-1-phophate receptor 1 (S1PR1)-ML056, corticotropin-releasing factor receptor1 (CRFR1)-CP-376395, Smoothened (Smo)-LY2940680, Smoothened (Smo)-SANT-1. For each complex, the PDB ID of the X-ray structure is indicated in brackets

(corticotropin-releasing factor receptor1 which is very located quite deep and almost vicinal to the intracellular membrane boundary (Fig. 6).

The recently solved X-ray structures of hSmo in complex with two antagonists (LY1940680, SANT-1) illustrate the fine plasticity of the receptor whose TM cavity adapts to its ligand. An explanation for this fine-tuning arises from the rotameric state of a single side chain (Leu325$^{3.36}$) which acts as a gatekeeper to close or open a back pocket in the TM crevice (Fig. 7).

When Leu325 side chain is perpendicular to the 7-TM main axis (e.g., when bound to LY2940680), the cavity stops at the middle of the helical bundle, therefore forcing the bound antagonist to mainly interact with ECL residues (Fig. 8). Conversely, when Leu325 side chain is parallel to the 7-TM main axis (e.g., bound to SANT-1), the back pocket is open and the antagonists deeply penetrate into a bigger cavity (800 Å^3 for SANT-1 bound Smo vs. 700 Å^3 for LY2940680-bound Smo).

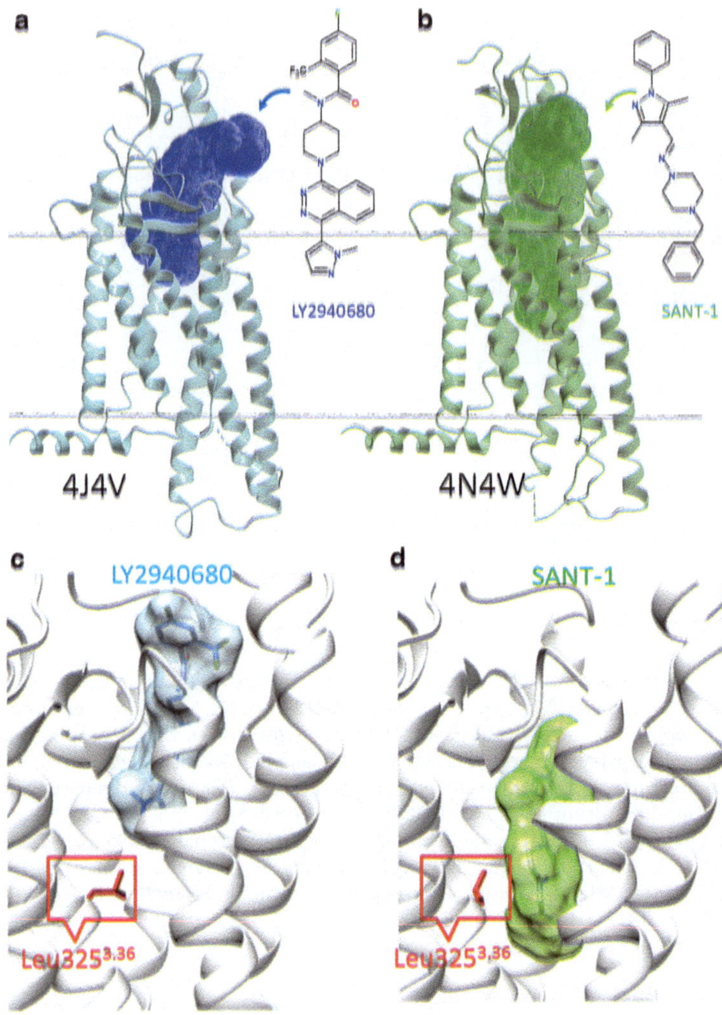

Fig. 7 X-ray structure of human Smo in complex with LY2940680 (**a**) and SANT-1 (**b**). The cavity to which the ligand binds is displayed by *blue* and *green meshes*, respectively. A close-up to the ligand-binding site pinpoints the crucial role of Leu325 whose rotameric state influences the volume of the TM cavity and the level of ligand penetration in the crevice

The flexibility of the receptor explains why Smo antagonists exhibit quite different chemotypes and structure–activity relationships since they most likely bind to a unique cavity but with partially overlapping residues. Mapping known site directed mutagenesis effects [7, 45, 47] to these recent structural data suggests that at least two kinds of TM-directed antagonists co-exist: (1) those binding mostly to the extracellular loops (LY2940680, cyclopamine, vismodegib; LDE225), (2) those penetrating deeply in the TM cavity (e.g., SANT-1). Further investigations are required to confirm the existence of antagonists bridging the two binding sites

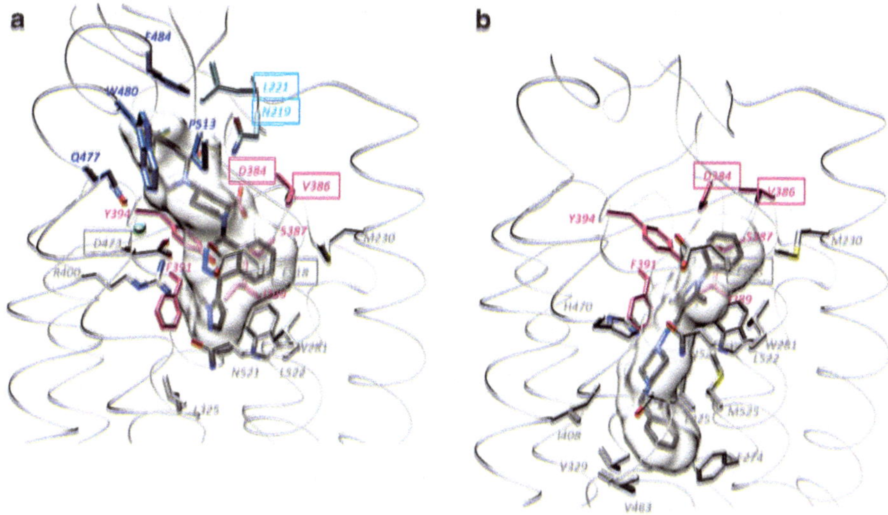

Fig. 8 Comparative binding mode of two antagonists (**a**, LY2940680; **b**, SANT-1) to human Smo. Only Smo side chains in close contact with the antagonist are displayed and labeled according to their position (*gray*, TM helices; *cyan*, ECD linker, *pink*, ECL2; *blue*, ECL3). *Boxed labels* indicate Smo residues which are mutated in medulloblastoma escaping chemotherapy with TM-directed Smo antagonists [7, 45, 46]

(Fig. 9) described above and occupying the entire TM cavity, and whether these different binding modes afford clinical advantages in blocking Hh signaling.

Up to now, all structural data on the TM domain have been gathered from the Smo inactive state. Agonist binding to GPCRs is known to promote a drastic conformational change of the 7-TM bundle [48]. Although Smo lacks the motifs at TM3 and TM6 engaged in the ionic lock that is released upon activation, it is likely that the Smo active state differs from that of the crystallized conformation, changes being minor at the binding site and amplified to the intracellular side of the TMs. Potent Smo agonists (SAG [49], GSA-10 [50]) have been described and proposed to bind to the TM cavity. Interestingly, GSA-10 effects on cell differentiation can be blocked by some (SANT-1, MRT-83, LDE225) but not all Smo antagonists (cyclopamine, GDC-0449, CUR61414) [50]. Conversely, SAG effects can be completely blocked by cyclopamine [49]. It is therefore likely that all agonists also bind to the same cavity but, as for antagonists, use non-overlapping sets of residues to anchor to the receptor. To support this hypothesis, minor chemical modifications of the SAG agonist led to Smo antagonists [51]. The functional consequences of part/full TM cavity occupation still have to be rationalized. It certainly affects the overall conformational flexibility of the receptor and its downstream capacity to induce G protein-dependent and/or independent signaling.

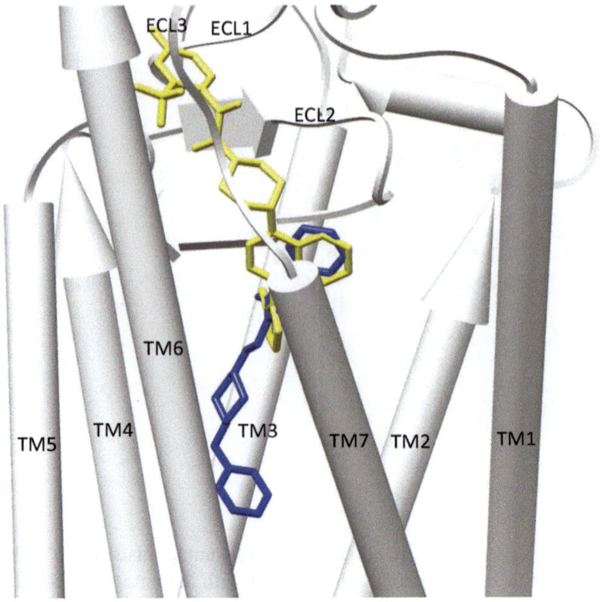

Fig. 9 Binding mode of LY2940680 (*yellow sticks*) and of SANT-1 (*blue sticks*) to human Smo (*gray ribbons*)

3.3 Smo Oligomeric State

Multiple evidences suggest that the functional unit of Smo may be a homodimer. First, Smo TM domain has been crystallized as a parallel dimer with a dimeric interface involving TMs 4 and 5 [8]. Interestingly, it is the only dimeric parallel interface among all crystallized GPCRs with a clear biological meaning with respect to residue conservation and packing [52]. Second, Hh-induced phosphorylation of Smo cytoplasmic tail by GRK2 induces a conformational switch of the receptor along with its dimerization [18, 43]. Third, a recent report evidenced that Smo form higher order oligomers in lipid rafts to enable Hh activity transduction [53]. Last, the ECD of the related Fz8 receptor exhibits an asymmetric dimeric interface when bound to its Wnt8 ligand. In addition to the above described ECD and TM ligand-binding sites, it is conceivable that additional sites exist on the surface of the Smo receptor, notably at the interface of the two interacting protomers. Targeting class A GPCR homo-and heterodimers has already led to dimer-biased ligands with unique pharmacological properties [54, 55]. It might be worth using the Smo dimer structure as a guide to design a novel generation of Smo modulators. Fortunately, the X-ray dimeric interface offers this opportunity by targeting one of the several drug-like cavities along the interface (Fig. 10).

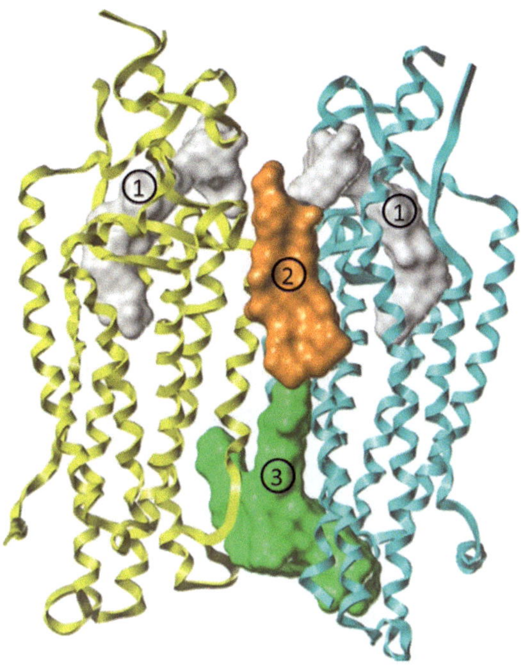

Fig. 10 X-ray structure of the Smo TM homodimer (monomer A, *yellow ribbons*; monomer B, *cyan ribbons*). The interface engages transmembrane domains (TM) 4 and 5 of both monomers. The dimer exhibits four structurally druggable cavities identified by the VolSite program [56]. The canonical antagonist pocket (1) is present in each monomer but two additional cavities (2, 3) are present at the dimer interface and constitute putative binding sites for a novel generation of Smo modulators

4 Conclusions

Although no endogenous Smo ligand has yet been identified, the receptor presents at least two well-characterized ligand-binding pockets: one in the extracellular domain (ECD), and one in the TM domain. This diversity of pockets explains the chemical diversity of Smo modulators and their different pharmacological properties (Fig. 11).

TM-binding agonists/antagonists have been well characterized and are likely to bind the same TM cavity albeit using different or overlapping residues. The recent X-ray structure of the TM domain in complex with three antagonists highlights overlapping binding modes, due to the rotameric state of a single residue controlling the depth of the TM cavity in which the ligand penetrates.

In addition to TM-directed ligands, we now know that sterols (oxy-, aza-sterols, glucocorticoids) are able to regulate Hh-induced Smo function upon binding to its ECD. Although they all bind to the same hydrophobic cleft, these compounds act either as positive or negative allosteric modulators of Hh signaling probably by stabilizing different quaternary structures of the Smo receptor. Interestingly, cyclopamine, an atypical TM antagonist able to accumulate Smo in cilia, is able to bind to both domains albeit with different affinities. Nachtergaele et al. [10] proposed that cyclopamine could be involved in a "hand-off" interaction between the ECD and the TM analogous to the manner by which cholesterol is transferred

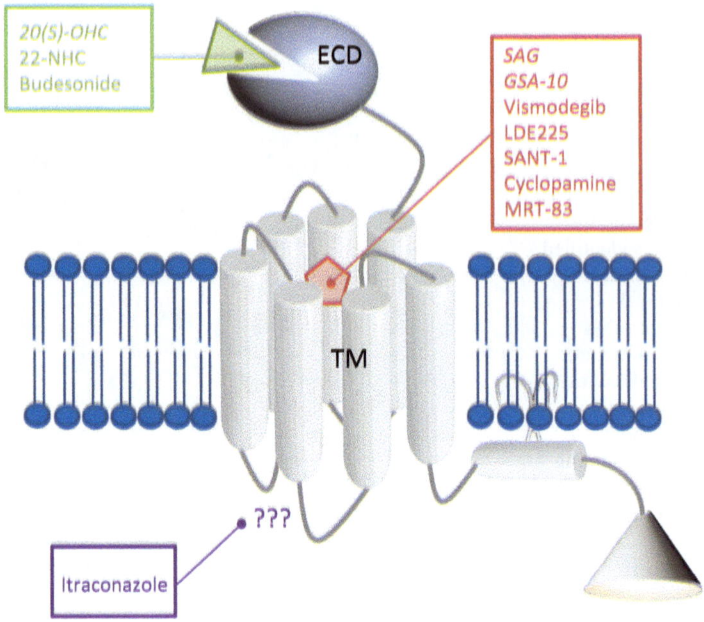

Fig. 11 Multiple sites for modulating Hh-induced Smo signaling at the transmembrane (TM) or the extracellular cysteine-rich domain (ECD). Known Smo agonists are indicated in italic. The binding site of the inverse agonist itraconazole [57] is still unknown

between NPC2 and NPC1 [58, 59]. This hypothesis is very interesting since NPC1 possesses in its sterol-binding domain the three core helices α1, α2, and α3 common to the Fz-like ECD family members (Fig. 4). It becomes now clear that the multiplicity of these binding modes provides an early explanation for the different functional effects observed with Smo antagonists. X-ray structure elucidation of Smo in a variety of states (active, inactive) and in complex with various ligands (ECD-binding sterols, TM-binding synthetic ligands) will enable the rational design of biased ligands with unique signaling properties and better tolerance to already observed antagonist-driven TM mutations.

Last remains the still unsolved and puzzling issue of a putative endogenous ligand, which under the control of Patched 1, may up- or down-regulate Smo activity by binding to the ECD or the TM domain (or both). Very few endogenous allosteric GPCR ligands [60] have been discovered up to now for the main reason that they have not been systematically searched. Currently available fast in vitro screening assays for endogenous metabolites [46] able to potentiate known Smo agonists may provide in a near future a clue to solve this mystery.

References

1. Ingham PW (2001) Hedgehog signaling: a tale of two lipids. Science 294:1879–1881
2. Scales SJ, de Sauvage FJ (2009) Mechanisms of Hedgehog pathway activation in cancer and implications for therapy. Trends Pharmacol Sci 30:303–312
3. Lagerstrom MC, Schioth HB (2008) Structural diversity of G protein-coupled receptors and significance for drug discovery. Nat Rev Drug Discov 7:339–357
4. Rudin CM (2012) Vismodegib. Clin Cancer Res 18:3218–3222
5. Amakye D, Jagani Z, Dorsch M (2013) Unraveling the therapeutic potential of the Hedgehog pathway in cancer. Nat Med 19:1410–1422
6. Ruat M, Hoch L, Faure H, Rognan D (2014) Targeting of Smoothened for therapeutic gain. Trends Pharmacol Sci 35:237–246
7. Yauch RL, Dijkgraaf GJ, Alicke B, Januario T, Ahn CP, Holcomb T, Pujara K, Stinson J, Callahan CA, Tang T, Bazan JF, Kan Z, Seshagiri S, Hann CL, Gould SE, Low JA, Rudin CM, de Sauvage FJ (2009) Smoothened mutation confers resistance to a Hedgehog pathway inhibitor in medulloblastoma. Science 326:572–574
8. Wang C, Wu H, Katritch V, Han GW, Huang XP, Liu W, Siu FY, Roth BL, Cherezov V, Stevens RC (2013) Structure of the human smoothened receptor bound to an antitumour agent. Nature 497:338–343
9. Rana R, Carroll CE, Lee HJ, Bao J, Marada S, Grace CR, Guibao CD, Ogden SK, Zheng JJ (2013) Structural insights into the role of the Smoothened cysteine-rich domain in Hedgehog signalling. Nat Commun 4:2965
10. Nachtergaele S, Whalen DM, Mydock LK, Zhao Z, Malinauskas T, Krishnan K, Ingham PW, Covey DF, Siebold C, Rohatgi R (2013) Structure and function of the Smoothened extracellular domain in vertebrate Hedgehog signaling. eLife 2:e01340
11. Wang C, Wu H, Han GW, Cherezov V, Stevens RC (2014) Structure of the human smoothened receptor in complex with SANT-1
12. Weierstall U, James D, Wang C, White TA, Wang D, Liu W, Spence JC, Bruce Doak R, Nelson G, Fromme P, Fromme R, Grotjohann I, Kupitz C, Zatsepin NA, Liu H, Basu S, Wacker D, Han GW, Katritch V, Boutet S, Messerschmidt M, Williams GJ, Koglin JE, Marvin Seibert M, Klinker M, Gati C, Shoeman RL, Barty A, Chapman HN, Kirian RA, Beyerlein KR, Stevens RC, Li D, Shah ST, Howe N, Caffrey M, Cherezov V (2014) Lipidic cubic phase injector facilitates membrane protein serial femtosecond crystallography. Nat Commun 5:3309
13. Riobo NA, Saucy B, Dilizio C, Manning DR (2006) Activation of heterotrimeric G proteins by Smoothened. Proc Natl Acad Sci U S A 103:12607–12612
14. Fredriksson R, Lagerstrom MC, Lundin LG, Schioth HB (2003) The G-protein-coupled receptors in the human genome form five main families. Phylogenetic analysis, paralogon groups, and fingerprints. Mol Pharmacol 63:1256–1272
15. The GPCR Network Homepage (2014) http://gpcr.scripps.edu/. Accessed Feb 2014
16. Dann CE, Hsieh JC, Rattner A, Sharma D, Nathans J, Leahy DJ (2001) Insights into Wnt binding and signalling from the structures of two Frizzled cysteine-rich domains. Nature 412:86–90
17. Janda CY, Waghray D, Levin AM, Thomas C, Garcia KC (2012) Structural basis of Wnt recognition by Frizzled. Science 337:59–64
18. Zhao Y, Tong C, Jiang J (2007) Hedgehog regulates smoothened activity by inducing a conformational switch. Nature 450:252–258
19. Aanstad P, Santos N, Corbit KC, Scherz PJ, le Trinh A, Salvenmoser W, Huisken J, Reiter JF, Stainier DY (2009) The extracellular domain of Smoothened regulates ciliary localization and is required for high-level Hh signaling. Curr Biol 19:1034–1039
20. Nedelcu D, Liu J, Xu Y, Jao C, Salic A (2013) Oxysterol binding to the extracellular domain of Smoothened in Hedgehog signaling. Nat Chem Biol 9:557–564

21. Myers BR, Sever N, Chong YC, Kim J, Belani JD, Rychnovsky S, Bazan JF, Beachy PA (2013) Hedgehog pathway modulation by multiple lipid binding sites on the smoothened effector of signal response. Dev Cell 26:346–357
22. Cooper MK, Wassif CA, Krakowiak PA, Taipale J, Gong R, Kelley RI, Porter FD, Beachy PA (2003) A defective response to Hedgehog signaling in disorders of cholesterol biosynthesis. Nat Genet 33:508–513
23. Corcoran RB, Scott MP (2006) Oxysterols stimulate Sonic hedgehog signal transduction and proliferation of medulloblastoma cells. Proc Natl Acad Sci U S A 103:8408–8413
24. Bidet M, Joubert O, Lacombe B, Ciantar M, Nehme R, Mollat P, Bretillon L, Faure H, Bittman R, Ruat M, Mus-Veteau I (2011) The hedgehog receptor patched is involved in cholesterol transport. PLoS One 6:e23834
25. Dwyer JR, Sever N, Carlson M, Nelson SF, Beachy PA, Parhami F (2007) Oxysterols are novel activators of the hedgehog signaling pathway in pluripotent mesenchymal cells. J Biol Chem 282:8959–8968
26. Nachtergaele S, Mydock LK, Krishnan K, Rammohan J, Schlesinger PH, Covey DF, Rohatgi R (2012) Oxysterols are allosteric activators of the oncoprotein Smoothened. Nat Chem Biol 8:211–220
27. Wang Y, Davidow L, Arvanites AC, Blanchard J, Lam K, Xu K, Oza V, Yoo JW, Ng JM, Curran T, Rubin LL, McMahon AP (2012) Glucocorticoid compounds modify smoothened localization and hedgehog pathway activity. Chem Biol 19:972–982
28. Taipale J, Cooper MK, Maiti T, Beachy PA (2002) Patched acts catalytically to suppress the activity of Smoothened. Nature 418:892–897
29. Eaton S (2008) Multiple roles for lipids in the Hedgehog signalling pathway. Nat Rev 9:437–445
30. Roberg-Larsen H, Strand MF, Grimsmo A, Olsen PA, Dembinski JL, Rise F, Lundanes E, Greibrokk T, Krauss S, Wilson SR (2012) High sensitivity measurements of active oxysterols with automated filtration/filter backflush-solid phase extraction-liquid chromatography-mass spectrometry. J Chromatogr A 1255:291–297
31. Liu W, Wacker D, Gati C, Han GW, James D, Wang D, Nelson G, Weierstall U, Katritch V, Barty A, Zatsepin NA, Li D, Messerschmidt M, Boutet S, Williams GJ, Koglin JE, Seibert MM, Wang C, Shah ST, Basu S, Fromme R, Kupitz C, Rendek KN, Grotjohann I, Fromme P, Kirian RA, Beyerlein KR, White TA, Chapman HN, Caffrey M, Spence JC, Stevens RC, Cherezov V (2013) Serial femtosecond crystallography of G protein-coupled receptors. Science 342:1521–1524
32. Palczewski K, Kumasaka T, Hori T, Behnke CA, Motoshima H, Fox BA, Le Trong I, Teller DC, Okada T, Stenkamp RE, Yamamoto M, Miyano M (2000) Crystal structure of rhodopsin: a G protein-coupled receptor. Science 289:739–745
33. The GPCR Structural Database (2014) http://zhanglab.ccmb.med.umich.edu/GPCRSD/. Accessed Jan 2014
34. Mus-Veteau I, Zito F, Demange P (2014) Membrane protein production for structural analysis. In: Mus-Veteau I (eds) Membrane protein production for structural analysis. Springer, New York
35. Ahn KH, Nishiyama A, Mierke DF, Kendall DA (2010) Hydrophobic residues in helix 8 of cannabinoid receptor 1 are critical for structural and functional properties. Biochemistry 49:502–511
36. Delos Santos NM, Gardner LA, White SW, Bahouth SW (2006) Characterization of the residues in helix 8 of the human beta1-adrenergic receptor that are involved in coupling the receptor to G proteins. J Biol Chem 281:12896–12907
37. Awwad HO, Millman EE, Alpizar-Foster E, Moore RH, Knoll BJ (2010) Mutating the dileucine motif of the human beta(2)-adrenoceptor reduces the high initial rate of receptor phosphorylation by GRK without affecting postendocytic sorting. Eur J Pharmacol 635:9–15

38. Kirchberg K, Kim TY, Moller M, Skegro D, Dasara Raju G, Granzin J, Buldt G, Schlesinger R, Alexiev U (2011) Conformational dynamics of helix 8 in the GPCR rhodopsin controls arrestin activation in the desensitization process. Proc Natl Acad Sci U S A 108:18690–18695
39. Aratake Y, Okuno T, Matsunobu T, Saeki K, Takayanagi R, Furuya S, Yokomizo T (2012) Helix 8 of leukotriene B4 receptor 1 inhibits ligand-induced internalization. FASEB J 26:4068–4078
40. Venkatakrishnan AJ, Deupi X, Lebon G, Tate CG, Schertler GF, Babu MM (2013) Molecular signatures of G-protein-coupled receptors. Nature 494:185–194
41. Xie J, Murone M, Luoh SM, Ryan A, Gu Q, Zhang C, Bonifas JM, Lam CW, Hynes M, Goddard A, Rosenthal A, Epstein EH Jr, de Sauvage FJ (1998) Activating Smoothened mutations in sporadic basal-cell carcinoma. Nature 391:90–92
42. Umbhauer M, Djiane A, Goisset C, Penzo-Mendez A, Riou JF, Boucaut JC, Shi DL (2000) The C-terminal cytoplasmic Lys-thr-X-X-X-Trp motif in frizzled receptors mediates Wnt/beta-catenin signalling. EMBO J 19:4944–4954
43. Chen Y, Li S, Tong C, Zhao Y, Wang B, Liu Y, Jia J, Jiang J (2010) G protein-coupled receptor kinase 2 promotes high-level Hedgehog signaling by regulating the active state of Smo through kinase-dependent and kinase-independent mechanisms in Drosophila. Genes Dev 24:2054–2067
44. Chen JK, Taipale J, Cooper MK, Beachy PA (2002) Inhibition of Hedgehog signaling by direct binding of cyclopamine to Smoothened. Genes Dev 16:2743–2748
45. Buonamici S, Williams J, Morrissey M, Wang A, Guo R, Vattay A, Hsiao K, Yuan J, Green J, Ospina B, Yu Q, Ostrom L, Fordjour P, Anderson DL, Monahan JE, Kelleher JF, Peukert S, Pan S, Wu X, Maira SM, Garcia-Echeverria C, Briggs KJ, Watkins DN, Yao YM, Lengauer C, Warmuth M, Sellers WR, Dorsch M (2010) Interfering with resistance to smoothened antagonists by inhibition of the PI3K pathway in medulloblastoma. Sci Transl Med 2:51ra70
46. Roberg-Larsen H, Strand MF, Krauss S, Wilson SR (2014) Metabolites in vertebrate Hedgehog signaling. Biochem Biophys Res Commun. doi:10.1016/j.bbrc.2014.01.087, http://dx.doi.org/10.1016/j.bbrc.2014.01.087
47. Dijkgraaf GJ, Alicke B, Weinmann L, Januario T, West K, Modrusan Z, Burdick D, Goldsmith R, Robarge K, Sutherlin D, Scales SJ, Gould SE, Yauch RL, de Sauvage FJ (2011) Small molecule inhibition of GDC-0449 refractory smoothened mutants and downstream mechanisms of drug resistance. Cancer Res 71:435–444
48. Rasmussen SG, DeVree BT, Zou Y, Kruse AC, Chung KY, Kobilka TS, Thian FS, Chae PS, Pardon E, Calinski D, Mathiesen JM, Shah ST, Lyons JA, Caffrey M, Gellman SH, Steyaert J, Skiniotis G, Weis WI, Sunahara RK, Kobilka BK (2011) Crystal structure of the beta2 adrenergic receptor-Gs protein complex. Nature 477:549–555
49. Chen JK, Taipale J, Young KE, Maiti T, Beachy PA (2002) Small molecule modulation of Smoothened activity. Proc Natl Acad Sci U S A 99:14071–14076
50. Gorojankina T, Hoch L, Faure H, Roudaut H, Traiffort E, Schoenfelder A, Girard N, Mann A, Manetti F, Solinas A, Petricci E, Taddei M, Ruat M (2013) Discovery, molecular and pharmacological characterization of GSA-10, a novel small-molecule positive modulator of Smoothened. Mol Pharmacol 83:1020–1029
51. Yang H, Xiang J, Wang N, Zhao Y, Hyman J, Li S, Jiang J, Chen JK, Yang Z, Lin S (2009) Converse conformational control of smoothened activity by structurally related small molecules. J Biol Chem 284:20876–20884
52. Duarte JM, Biyani N, Baskaran K, Capitani G (2013) An analysis of oligomerization interfaces in transmembrane proteins. BMC Struct Biol 13:21
53. Shi D, Lv X, Zhang Z, Yang X, Zhou Z, Zhang L, Zhao Y (2013) Smoothened oligomerization/higher order clustering in lipid rafts is essential for high Hedgehog activity transduction. J Biol Chem 288:12605–12614
54. Gomes I, Fujita W, Gupta A, Saldanha SA, Negri A, Pinello CE, Eberhart C, Roberts E, Filizola M, Hodder P, Devi LA (2013) Identification of a mu-delta opioid receptor heteromer-biased agonist with antinociceptive activity. Proc Natl Acad Sci U S A 110:12072–12077

55. Hiller C, Kuhhorn J, Gmeiner P (2013) Class A G-protein-coupled receptor (GPCR) dimers and bivalent ligands. J Med Chem 56:6542–6559
56. Desaphy J, Azdimousa K, Kellenberger E, Rognan D (2012) Comparison and druggability prediction of protein-ligand binding sites from pharmacophore-annotated cavity shapes. J Chem Inf Model 52:2287–2299
57. Kim J, Tang JY, Gong R, Kim J, Lee JJ, Clemons KV, Chong CR, Chang KS, Fereshteh M, Gardner D, Reya T, Liu JO, Epstein EH, Stevens DA, Beachy PA (2010) Itraconazole, a commonly used antifungal that inhibits Hedgehog pathway activity and cancer growth. Cancer Cell 17:388–399
58. Kwon HJ, Abi-Mosleh L, Wang ML, Deisenhofer J, Goldstein JL, Brown MS, Infante RE (2009) Structure of N-terminal domain of NPC1 reveals distinct subdomains for binding and transfer of cholesterol. Cell 137:1213–1224
59. Wang ML, Motamed M, Infante RE, Abi-Mosleh L, Kwon HJ, Brown MS, Goldstein JL (2010) Identification of surface residues on Niemann-Pick C2 essential for hydrophobic handoff of cholesterol to NPC1 in lysosomes. Cell Metab 12:166–173
60. Pamplona FA, Ferreira J, de Lima M, Jr O, Duarte FS, Bento AF, Forner S, Villarinho JG, Bellocchio L, Wotjak CT, Lerner R, Monory K, Lutz B, Canetti C, Matias I, Calixto JB, Marsicano G, Guimaraes MZ, Takahashi RN (2012) Anti-inflammatory lipoxin A4 is an endogenous allosteric enhancer of CB1 cannabinoid receptor. Proc Natl Acad Sci U S A 109:21134–21139

Top Med Chem (2015) 16: 147–172
DOI: 10.1007/7355_2014_83

Published online: 20 January 2015

Smoothened, Stem Cell Maintenance and Brain Diseases

Martial Ruat, Hélène Faure, and Mathieu Daynac

Abstract The Smoothened (Smo) receptor is a key transducer of the Sonic Hedgehog (Shh) signaling pathway in the brain. Recent studies in rodents have highlighted its major role in the maintenance of neural stem and progenitor cells in the two main neurogenic niches of the adult brain: the subventricular zone of the lateral ventricles and the subgranular zone of the dentate gyrus in the hippocampus. Smo may also regulate brain responses to various injuries, and its modulation in the primary cilia of brain cells is essential for regulating Shh signals. Recent clinical trials have underlined the therapeutic value of some Smo antagonists for the treatment of Hedgehog-linked medulloblastomas. Here, we review recent findings on the roles of Smo in the adult brain, and unravel research on the clinical implications for the treatment of brain diseases, that are increasingly under investigation.

Keywords Astrocytes, Clinical trials, Hedgehog, Medulloblastoma, Patched

Contents

M. Ruat (✉), H. Faure and M. Daynac
Molecules and Circuits Department, Signal Transduction and Developmental Neuropharmacology Team, NeuroPSI Institute, CNRS, UMR 9197, 1 Avenue de la Terrasse, 91198 Gif-sur-Yvette, France
e-mail: ruat@inaf.cnrs-gif.fr

Abbreviations

aNSC	Activated neural stem cell
AD	Alzheimer's disease
APP	Amyloid precursor protein
CNS	Central nervous system
DG	Dentate Gyrus
Dhh	Desert Hedgehog
DS	Down syndrome
EGFR	Epidermal growth factor receptor
Gas1	Growth arrest-specific 1 protein
GCL	Granular cell layer
GCP	Cerebellar granule cell precursor
GDNF	Glial cell line-derived neurotrophic factor
GFAP	Glial fibrillary acidic protein
GLAST	Astrocyte-specific glutamate transporter
GPCR	G-protein-coupled receptor
Hh	Hedgehog
HIP	Hedgehog-interacting protein
Ihh	Indian Hedgehog
JS	Joubert syndrome
LPC	Lysophosphatidyl choline
LV	Lateral ventricle
MKS	Meckel syndrome
NG2	Neuron-glial antigen 2
NMDA	*N*-methyl-D-aspartate
NSC	Neural stem cell
OB	Olfactory bulb
OL	Oligodendrocyte
OPC	Oligodendrocyte precursor cell
PlGF	Placental growth factor
Ptc	Patched
qNSC	Quiescent neural stem cell

RMS	Rostral migratory stream
SGZ	Subgranular zone
Shh	Sonic Hedgehog
ShhN	Aminoterminal fragment of Shh
Smo	Smoothened
Sufu	Suppressor of fused
SVZ	Subventricular zone
TAC	Transit-amplifying cell

1 Introduction

The Hedgehog (Hh) signaling pathway is well characterized for its roles in the patterning and growth of brain structures during embryogenesis. Along with other morphogens implicated in stem cell biology, including those from the Wnt and Notch families, Hh peptides are essential regulators of embryonic organ development and adult tissue homeostasis. Under physiological or pathophysiological circumstances, Hh pathway is activated and is tightly regulated at the level of the stem cell niches that are present in many adult tissues, including the brain and the spinal cord [1–5]. The discovery of a specific Hh signaling pathway in the adult rodent brain [6] has generated considerable interest, while novel functions for Hh peptides have progressively emerged [7–9]. Moreover, the importance of Hh signaling in adult brain plasticity is demonstrated by its role in neural stem cell (NSC) maintenance in several adult neurogenic niches [10, 11]. Humans affected by the Gorlin syndrome have inactivating mutations in the Hh receptor Patched (Ptc), characterized as a negative regulator of Hh signaling. These patients have a susceptibility to medulloblastoma, one of the most malignant childhood brain tumors. As a consequence of these mutations, the Ptc-mediated inhibition exerted on the Smoothened (Smo) receptor, a G-protein-coupled receptor (GPCR), which is a key transducer of the Hh signaling pathway, is relieved initiating the tumorigenic process. Thus, inhibiting the Hh signaling pathway with small molecule inhibitors has generated considerable interest for treating these tumors, with several Smo inhibitors currently being evaluated for treating medulloblastomas [12–16]. The recent discovery that Hh signaling depends on primary cilia [17], an important signaling center, has fostered studies aimed at characterizing the distribution and the regulation of the pathway, including Smo [5, 18–20]. Our review focuses on the function of Smo in the brain, and the potential advantage of manipulating its activity for treating brain diseases.

2 Smo Is a Key Transducer of Hh Signals

2.1 Canonical Hh Signaling

Sonic Hh (Shh), Indian Hh (Ihh), and Desert Hh (Dhh) are morphogen molecules synthesized as precursors that undergo autocatalytic cleavage into N- and C-terminal domains. The N-terminal domain of these molecules (ShhN for Shh) is highly conserved among species and is responsible for the majority of their signaling activities. In contrast, these molecules have a divergent C-terminal domain. This domain is involved in the autoproteolysis reaction, and in the addition of a cholesterol moiety, which is covalently attached to the C-terminus end of the N-terminal domain. This cholesterol molecule participates in tethering the N-terminal domain to the plasma membrane and presumably restricts the tissue localization of Hh signaling. As demonstrated for ShhN, addition of another lipid to the N-terminal cysteine residue dramatically increases the biological activity of the protein [21]. Thus, these lipid modifications may exert an essential role in targeting the protein to lipid rafts in brain cells [22] that potentially represent functional platforms for Shh signaling [23].

Hh proteins mediate their action via a receptor complex consisting of two transmembrane proteins: Ptc, the Shh receptor, which displays a transporter-like structure, and Smo, a GPCR, which transduces the Shh signal downstream of Ptc. The repression exerted by Ptc on Smo is relieved when Shh binds Ptc, initiating a complex signaling cascade. This leads to the activation of transcription factors of the Gli family (Gli1-3) and to the transcription of target genes including *Ptc* and *Gli1* themselves (Fig. 1) [28]. Activation of the canonical Shh pathway inhibits the processing of the Gli transcription factor into its transcriptional repressor forms and leads to the concomitant accumulation of its activator forms. Gli1 constitutes a convenient readout for pathway activation and amplifies the Hh response. Gli2 and Gli3 function mainly as transcriptional activator and repressor, respectively, even if both can show opposite activity in specific contexts [28, 29].

Several additional proteins bind Hh proteins with high affinity, function as Shh coreceptors, and promote Hh signaling by unknown mechanisms [2, 30]. These include the negative regulator Hedgehog-interacting protein (Hip), found in soluble and membrane-associated forms in brain regions [23], the two cell surface immunoglobulin/fibronectin proteins, Boc and Cdo, and the growth arrest-specific 1 protein (Gas1). The structurally related Boc and Cdo are integral membrane proteins conserved from *Drosophila* to rodents, whereas Gas1, a glycosylphosphatidylinositol-anchored plasma membrane protein, is specific to Hh signaling in vertebrates. A proposed model is that Gas1, Cdo, and Boc form a physical complex with Ptc and function as essential coreceptors that mediate multiple cellular responses to Hh. However, this requirement of Hh coreceptors depends on the cell type and developmental stage, whereas their role on Smo regulation is less clearly understood [30, 31].

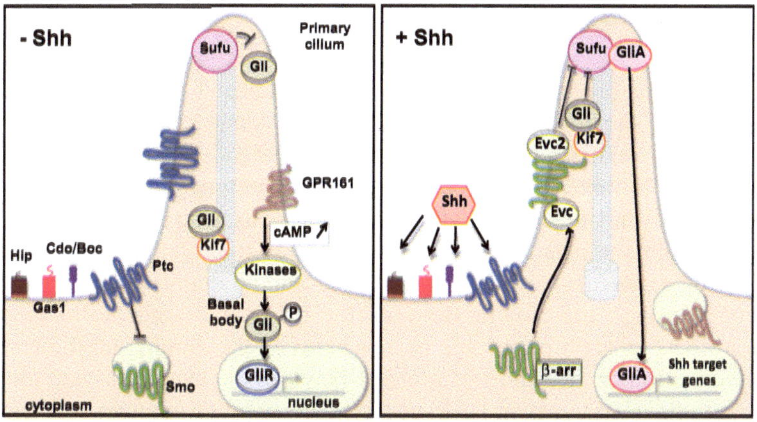

Fig. 1 Smoothened transduction of canonical Sonic Hedgehog signal at the primary cilium. In the absence of Sonic Hedgehog ligand (−Shh), the receptor Patched (Ptc), located in the cilium, inhibits by a yet unknown mechanism the 7-transmembrane receptor Smoothened (Smo), mostly found outside the cilium. Gli transcription factors (Gli) are in a complex with the anterograde kinesin-like motor protein Kif7. Gli truncation and phosphorylation are promoted by the repressor factors such as suppressor of fused (Sufu) and kinases including protein kinase A (PKA), casein kinase 1α and glycogen synthase kinase 3β. The constitutive activity of GPR161, another member of the G-protein-coupled receptor superfamily [24], induces an increase of cAMP at the primary cilia resulting in an activation of PKA leading to the accumulation of Gli repressors (GliR) and inhibition of Shh target genes. Hip, a negative regulator of the pathway found as both membrane-associated and soluble forms, and the membrane proteins Gas1, Cdo, and Boc that act as positive regulators bind Shh with high affinity. In the presence of Shh ligand (+Shh), Smo inhibition is relieved allowing its translocation and accumulation in the cilium, which involves interaction with β-arrestin (β-Arr) [25], and association with Evc/Evc2 complex [26] to transduce the signal intracellularly. This leads to Sufu inhibition and Gli conversion into their activated forms (GliA). GliA enters the nucleus and activates transcription of Shh target genes including Ptc and Gli1. Smo may move through a lateral transport pathway from the plasma membrane to the ciliary membrane [27]

2.2 Smo and Noncanonical Hh Signaling

Besides the canonical Hh pathway, noncanonical mechanisms of Hh signaling are also reported. Some of these result from Smo activation and might be unrelated to Gli-mediated transcription, but instead depend on non-transcriptional events. Examples include (1) proposed Src kinase activation, which mediates Hh effects on spinal commissural axonal projections toward the ventral midline [32], (2) synchronous Ca^{2+} spikes and inositol triphosphate oscillations at the neuronal primary cilium [33], (3) and glucose uptake through a Ca^{2+}-AMP-activated protein (AMPK), cilia-dependent pathway [34]. Furthermore, in agreement with Smo being a GPCR, it associates with heterotrimeric Gi proteins and stimulates small GTPases Rac1 and RhoA, leading to fibroblast migration [35]. These and other noncanonical mechanisms, involving Hh binding to Ptc or to other putative Hh receptors, or involving Smo activation, are recently reviewed elsewhere [36–39].

2.3 Smo Signals at the Primary Cilium

In vertebrates, the primary cilium was recently proposed as a major transduction platform for Hh signaling. The primary cilium is defined as a microtubule-based organelle, approximately 1–5 μm in length, which extends from the cell surface as a single, nonmotile, antenna-like structure and is present on most cell types in embryonic and adult tissues [19, 40]. The ciliary basal body is formed from the mother centriole and acts as a docking area for a large number of pericentriolar proteins. The axoneme, made up of nine microtubule doublets, extends from the basal body through the cilium. Between the basal body and the axoneme, the transition fibers create a permeable barrier between the cilium and the rest of the cell. Selective import or export of proteins between the cytoplasm and the cilium requires intraflagellar transport (IFT) particles forming two complexes B and A that use the anterograde kinesin-2 (also known as the Kif3 motor complex) and the retrograde dynein motors, respectively [41, 42]. The primary cilium differs from the secondary motile cilia. In the latter, the axoneme contains an extra central pair of microtubules linked to the nine outer microtubule pairs. Intense genetic, molecular, biochemical, and pharmacological studies were recently conducted to identify components of signal transduction pathways that are present in this organelle. Trafficking of Hh signaling proteins along the cilium in stem or precursor cells is vital during neural development, for several genetic diseases and in cancer [18, 19].

Smo, Ptc, Gli1-3, and the negative regulator Sufu were detected at the primary cilium [43–46]. Shh was identified close to the cilium base in target neural progenitors, during active Shh signaling in the neural tube [47]. Ptc is believed to be localized to the base of the cilium in the absence of its ligand and is proposed to inhibit signaling by preventing Smo localization to the cilium (Fig. 1). Upon ligand binding, Ptc is removed, and Smo is simultaneously localized to the cilium. β-Arrestins might mediate Smo interaction with Kif3a kinesin motor protein, regulating Smo localization to primary cilia [25, 46]. Alternatively, Smo may move via a lateral transport pathway from the plasma membrane to the ciliary membrane [27]. Finally, phosphorylation of mouse Smo carboxyl terminal tail by the serine/threonine kinases GIRK2 and CK1α was found to activate the receptor and promote its ciliary accumulation [48]. Activation of Smo would then antagonize the activity of Sufu, which negatively regulates Gli transcription factors (Fig. 1). The kinesin Kif7, the vertebrate homologue of *Drosophila* Costal2, regulating the activity of Cubitus interruptus (the homologue of Gli transcription factors) [49, 50], is proposed to play a major role for tethering the components of the Hh pathway at the primary cilium. Kif7 motor domain retains the sequences characteristic of kinesin motors suggesting it may behave as an anterograde motor protein. Kif7 would act downstream of Smo and upstream of Gli2. In the absence of Shh, Kif7 is localized to the cilium base where it forms a complex in particular with Gli proteins and promotes processing of Gli repressor forms. Upon ligand stimulation, Kif7 translocates to the cilium tip and would block Sufu, resulting in the accumulation and activation of the Gli proteins [51, 52]. Interestingly, both Gli2 and Smo were found to require the retrograde motor dynein for exiting the cilium [53]. Thus, canonical Hh signaling requires complex trafficking of Hh components to the primary cilium.

3 Smo Regulates the Maintenance of Stem Cell Niches in the Adult Brain

Recent advances in adult neurogenesis highlight the capacity of the brain to generate new neurons throughout adult life. Stem cells were characterized in several regions of the adult brain including the subventricular zone (SVZ) of the lateral ventricles (LV) and the subgranular zone (SGZ) of the dentate gyrus (DG) within the hippocampus (Figs. 2, 3, and 4) [61, 62, 66, 67]. Both neurogenic niches contain astrocytes displaying features of slow-dividing adult NSCs that give rise to proliferating progenitors, primary precursors of migrating neuroblasts. The SVZ and the SGZ provide new neurons that are fully integrated into the granular and glomerular cell layers of the olfactory bulbs and into the granular cell layer of the DG in the hippocampus, respectively (Figs. 3a, b and 4). In defined pathological conditions, newly generated cells deviate from their normal destination and migrate toward the injured tissue [68]. Hippocampal neurogenesis is proposed to modulate multiple aspects of hippocampus-dependent roles, including long-term memory and mood [69]. Thus, drugs that affect signaling pathways regulating hippocampal neurogenesis may have implications in cognitive function and depressive symptoms.

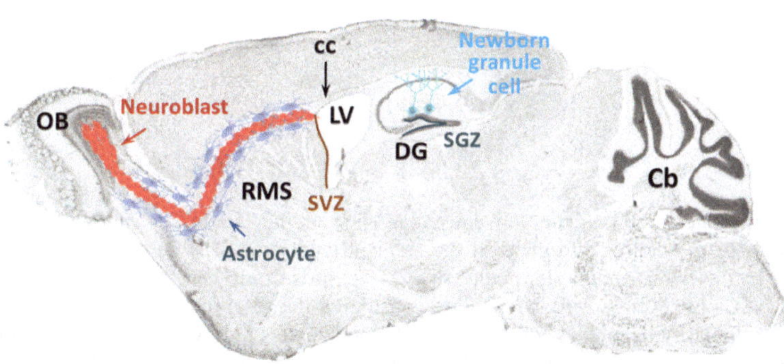

Fig. 2 Main neurogenic zones of the adult mice brain. Sagittal section of an adult mouse brain with the two main neurogenic regions: the subventricular zone (SVZ) of the lateral ventricles (LV) and the subgranular zone (SGZ) of dentate gyrus (DG) in the hippocampus. Neural stem cells (NSCs) from the SVZ produce neuronal precursors (neuroblasts) that migrate to the olfactory bulbs (OB) through the rostral migratory stream (RMS), a specialized migratory route surrounded by astrocytes. In the OB, neuroblasts differentiate into functional neurons that integrate local neural circuits involved in olfactory processing. NSCs from the SGZ produce functional neurons (granule cells) with a function in learning and memory. *Cb* cerebellum, *cc* corpus callosum

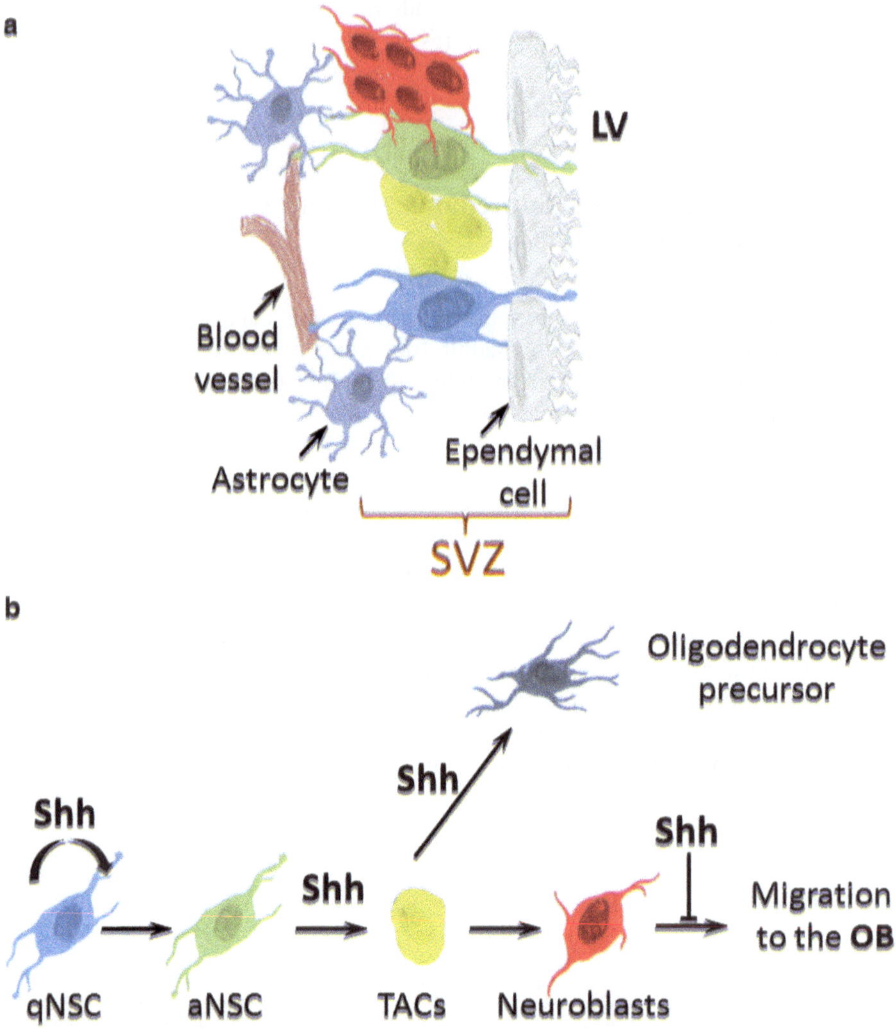

Fig. 3 Neural stem cells in the subventricular zone of the lateral ventricles respond to Shh. (**a**) Drawing of the neurogenic niche at the subventricular zone (SVZ) of the lateral ventricles (LV). (**b**) In the SVZ, neural stem cells (NSCs) comprise quiescent NSCs (qNSCs; *blue*) and activated NSCs (aNSCs; *green*). qNSCs are considered the reservoir for adult neurogenesis and have the particularity to contact both the LV and blood vessels [54]. Once they enter proliferation, they become aNSCs that give rise to proliferating transit-amplifying cells (TACs; *yellow*). TACs generate neuroblasts (*red*). Next, neuroblasts migrate as chains to the olfactory bulbs (OB) via the rostral migratory stream. In the OB, they differentiate into GABA- and dopamine-producing granule and periglomerular interneurons. Oligodendrocyte precursors are generated from a small fraction of TACs and migrate radially out of the SVZ to the corpus callosum and cortex [55, 56]. qNSCs and migrating neuroblasts respond to Shh [57–59]. Chemoattractivity of Shh blocks the exit of migrating neuroblasts to the OB [60]. Shh may also regulate the number of oligodendrocyte precursors derived from the SVZ [59]

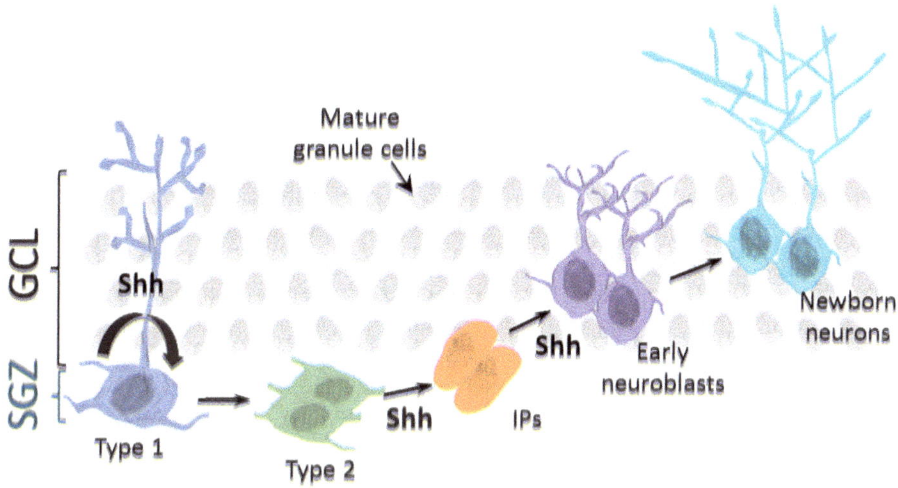

Fig. 4 Hippocampal neural stem cells in the subgranular zone of the dentate gyrus are regulated by Shh signaling. In the dentate gyrus (DG) of the hippocampus (see Fig. 2), radial glia NSCs (type 1) (*deep blue*) and nonradial (type 2) cells (*green*) are present in the subgranular zone (SGZ). Type 1 cells are slowly dividing cells that were proposed to give rise to type 2 cells by asymmetric division. Type 2 cells are actively proliferating and produce intermediate neural progenitors (IPs; *orange*) that in turn give rise to early immature neuroblasts (*purple*). The latest will become functional glutamatergic granule neurons fully integrated within the granular cell layer (GCL) [61, 62]. NSCs and progenitors in the SGZ respond to Shh signaling. Smo deletion in radial glial NSCs in the SGZ results in a disorganized dentate gyrus and reduced neurogenesis [63, 64]. Smo activation increases proliferation in the SGZ of the dentate gyrus [65]

3.1 *Smo and the Regulation of Stem and Progenitors Cells in the Subventricular Zone of the Lateral Ventricles*

The early identification of the distribution of Smo transcripts as well as the other components of the canonical pathway, such as Shh and Ptc, in the adult brain, was in support of a role of Hh signaling in germinal niches, including the SVZ [22, 57, 70–72]. The analysis of *Gli1-CreER*^{T2} reporter mice indicated that astrocyte-like NSCs are the primary Shh responsive cells in the SVZ [58]. These observations are further corroborated by studies of Gli transcription factors in several reporter mice [73, 74]. Smo is expressed in this region as shown by its immunolocalization in astrocyte-like stem cell population [73]. The direct injection of active Shh protein in both rat and mice LV was followed by a rapid *Ptc* transcription in the SVZ, which was the first argument supporting the hypothesis that this pathway is functional in the adult CNS [57]. Ptc transcription was blocked by the direct injection of a Smo antagonist in the LV, which indicates that Smo is crucial for Hh canonical signaling in this niche [75]. These data support the presence of Shh responsive cells within this germinal niche and the possibility that it might have roles other than the regulation of proliferative activity of precursor cells [57]. However, pharmacological activation of Hh signaling by administration of a Smo agonist in adult mice increases the number of proliferating cells in the SVZ and upregulates *Gli1* in these cells [76]. Conversely, the pharmacological blockade of Hh signaling by systemic injection of cyclopamine, a Smo antagonist, results in a decrease of *Gli1* and *Gli3* expression and in cell proliferation in the SVZ [77]. However, the in vivo effects of cyclopamine have been questioned in experiments where injection of the drug in the LV of adult rats failed to modulate the number of proliferating SVZ precursors [78]. Nevertheless, Shh protein added directly to neurospheres cultured from SVZ, not only increases their number, but also modulates stem cell proliferation and the balance between self-renewal and differentiation [65, 76, 77]. Altogether, these data suggest different modes of regulating Hh signaling in vitro and in vivo.

The removal of Shh signaling in mice with conditional null alleles of Smo was investigated using a transgenic mouse line expressing the Cre recombinase under the nestin promoter—a ubiquitous marker of neural progenitors activated by embryonic day E12.5. In these animals, Shh signaling is required for the maintenance of adult neural progenitor populations [76]. Both NSCs and transit-amplifying cells (TACs) are depleted postnatally. However, migrating neuroblasts expand dramatically, mostly fail to migrate to the olfactory bulb, and are depleted by postnatal day 30. Thus, Smo is required for the maintenance of NSCs and TACs and indirectly for the migration of precursor cells generated in the SVZ [79]. When Smo function is conditionally removed from the SVZ in adult mice brain using the tamoxifen-inducible transgenic nestin-Cre^{ERT2}, both proliferation and neurogenesis are modified and fail to recover. These observations indicate that Hh signaling through Smo activation is required for stem cell maintenance and TAC proliferation during adulthood in addition to the establishment of the adult stem cell niche [80]. Smo modulation in the SVZ may also contribute to the generation of multiple interneuron subtypes in the olfactory bulb [81]. Moreover, Shh chemoattraction for SVZ-derived neuroblasts presumably involves Smo modulation, since pharmacological blockade of the receptor antagonizes the Shh chemoattraction observed in vitro on SVZ-derived neuronal progenitors (Fig. 3b) [60].

Genetic activation of Hh signaling in adult NSCs was recently achieved through conditional Ptc inactivation in astrocyte-specific glutamate transporter (GLAST)-expressing cells. This activation presumably results from the suspension of the inhibitory effect that Ptc exerts on Smo and leads to NSC expansion and a depletion of their direct progeny. Thus, Smo activation in NSCs of the SVZ may modulate NSC symmetric division through a process involving the Notch signaling pathway [59]. Interestingly, activation of Hh signaling in NSCs increases the stemness of these cells without promoting tumor formation. Thus, Smo activation through genetic Ptc inactivation results in a blockade of epidermal growth factor receptor EGFR$^+$ proliferative cells. Importantly, Hh signaling activation in the SVZ of these mice influences the fate of SVZ-derived precursors toward either the oligodendroglial phenotype in the corpus callosum or the glutamatergic and dopaminergic phenotypes. Both glutamatergic and dopaminergic neuronal populations constitute the olfactory bulb periglomerular interneurons, the specification of which may be of high interest in the context of brain repair [82, 83]. Therefore, it will also be important to address if small molecule agonists or antagonists of Smo, including the oxysterol derivatives recently identified as potential endogenous Smo ligands [84–88], regulate NSC stemness in vivo.

3.2 Smo Is Essential for Hippocampal Stem Cell Maintenance

The discovery of Shh protein in various adult rodent brain regions, including the hippocampus [22], favored its potential role in regulating adult hippocampal neurogenesis. In agreement with its axonal transport [22], it was proposed that Shh originating from the subpallial septum region [70], via the septo-hippocampal pathway, may be the source of Shh-regulating hippocampal neurogenesis [65]. Whether hippocampal NSCs are also a source of Shh [89] needs further investigation since recent developmental studies indicated that Shh involved in the regulation of adult SGZ mainly originates from neurons [90].

Activation of Smo in the hippocampus following oral administration of a Smo agonist increases the number of proliferating cells induced in the SGZ. Conversely, treatment with the Smo inhibitor cyclopamine reduced hippocampal neural progenitor proliferation in vivo [65, 76, 78]. These data identified Smo, for the first time, as a key regulator of adult hippocampal NSCs. Smo signaling through the primary cilia was then proposed to play a critical role in the expansion and establishment of postnatal hippocampal progenitors. Mice lacking Smo in glial fibrillary acidic protein (GFAP)-positive neural precursor cells show a defect in hippocampal neurogenesis reflected by a very small DG, a markedly reduced proliferation and few newly generated neurons (Fig. 4) [63, 64]. This phenotype is close to the one observed in mice lacking the kinesin Kif3a [91] in GFAP-positive cells, which display defective ciliogenesis, impairment of Shh signaling in granule neuron precursors, and no generation of radial astrocytes that function as the SGZ stem cells [63]. Overexpression of an oncogenic form of Smo (SmoM2) in hippocampal granule neuron precursors does not result in tumor formation, but in a clear increase of granule neuron precursors [63]. In these cells, SmoM2 is localized to the primary cilia, in agreement with its previously found subcellular localization in fibroblasts [92]. Such an effect was not observed in the Kif3a mutant. Therefore, correct Smo signaling is necessary for hippocampal stem cell development. When adult neurogenesis is constantly decreased by conditional ablation of primary cilia in adult GFAP-positive NSCs, a delay in spatial learning and alteration of spatial novelty recognition was observed [93]. However, it is not yet known whether cognitive deficits observed in ciliopathies are related in some aspects to the alteration of Hh signaling through Smo regulation at the primary cilia.

In the stumpy mutant lacking ciliary axonemes in the hippocampus [64, 94], improper Gli3 processing was also associated with the loss of cilia. The thinner dentate granule cell layer observed in these animals was linked to a deficiency in NSCs and cell cycle exit, presumably reflecting a decrease of Hh signaling at the primary cilia. Among the Gli transcription factors, *Gli1* was found to be the only one that was transcriptionally induced following Shh pathway activation in the SGZ and was required for the self-renewal of postnatal NSCs [95]. In this area, *Gli1* expression is associated with the upregulation of genes not previously identified as Gli1 transcriptional targets, such as proapoptotic genes and genes governing the G2-M transition. This might be related to the unique ability of NSCs to maintain tight regulation over cell cycle progression. Although hippocampal development in embryos and NSC maintenance require Sox2-dependent regulation of Shh, this type of regulation remains to be determined in adult hippocampal NSCs [89].

Since hippocampal neurogenesis is closely associated with learning and memory, which are impaired in Alzheimer's disease (AD), it is questioned whether Shh signaling is impaired in the hippocampus of a mouse model of AD. APP23 mice are transgenic mice expressing mutated human β-amyloid precursor protein (βAPP) (Swedish double mutation, KM670/671NL) [96] and develop senile plaques in the hippocampus. In this widely used mouse model of AD, Shh signaling is deregulated in the hippocampus and is associated with the exit of NSC quiescence toward proliferation and differentiation, thus leading to a progressive exhaustion of the NSC pool. This results in a decrease in the production of new neurons [97]. More studies in other mouse models of AD are required to determine whether treatment of these mice with Smo modulators would modify the pattern of β-amyloid deposits.

The distribution of Smo protein in adult hippocampal mossy fibers [98, 99] and its proposed regulation by synaptic activity involving glutamatergic transmission suggest additional roles for Hh signaling in the control of hippocampal functions [78, 99]. Electroconvulsive seizures, a potent treatment for depressive disorders, induce a rapid and robust reduction in *Smo* transcription in the DG. Furthermore, research suggests that adult hippocampal neurogenesis is required for the behavioral activity of antidepressants. However, further investigations are required to prove the effect of Smo modulation on antidepressant activity [5].

4 Smo Regulation of Cerebellar Precursors and Tumor Stem Cells

4.1 Hh-Dependent Formation of Medulloblastoma

Abnormal proliferation of cerebellar granule cell precursors (GCPs) is associated with the development of medulloblastoma. These growing tumors of the cerebellum are particularly aggressive in children and are divided in distinct molecular subgroups, which include the WNT and Shh tumors [100]. The Shh tumors are characterized by Shh pathway activation and have a poor prognosis. The recent approval of the Smo antagonist GDC-0449 for the treatment of metastatic and locally advanced basal cell carcinoma has given new hope for the treatment of these tumors despite important drawbacks such as treatment resistance. These tumors were initially considered as Hh ligand independent, since aberrant signaling activation is linked to pathway-activating mutations in Ptc, Sufu, or Smo [101]. Humans affected by the Gorlin syndrome display a higher incidence of medulloblastomas. The Gorlin syndrome, an autosomal dominant inherited disorder, is characterized by Ptc-inactivating mutations resulting in a high number of basal cell carcinoma and with facial and skeletal anomalies. Gene silencing induced by methylation of the Ptc promoter may be responsible for tumor formation [102]. However, the insulin-like growth factor pathway and several other signals are presumably involved in regulating Hh-dependent formation of medulloblastomas as shown in Ptc heterozygous mice [103, 104]. Brain tumor stem cells derived from medulloblastomas occurring in these mice have a markedly higher basal Gli1 expression than normal stem cells. They do not undergo apoptosis, and their proliferation rate is higher. This suggests a deficit in the protective mechanisms that lead nonmalignant stem cells to restrain hyperproliferation in the context of potentially transforming mitogenic signals [95]. Conditional *Ptc* deletion, in GFAP-expressing NSCs, also promoted medulloblastoma in adults, only after NSCs are committed to granule neural precursors in the developing cerebellum. This suggests that this tumor is associated with neuronal lineage commitment [105].

Recently, a progenitor cell population expressing Nestin, a marker for multipotent cells, was identified in the deep part of the external germinative layer of the cerebellum and was also proposed to be the origin for medulloblastomas [106]. These precursor cells are quiescent, respond to Shh signaling, and formed tumors that might be related to genomic instability. Further studies are needed to investigate the effects of Smo inhibitors on the treatment of these medulloblastomas and to characterize resistances, if observed [107].

Shh secreted by Purkinje cells of the cerebellum were proposed to act as a mitogen on GCPs that express both Ptc and Smo and to play an important role for the development of medulloblastoma [70, 108–110]. However, tumor-derived Shh was also proposed to induce the production of placental growth factor (PlGF) from the cerebellar stroma leading to activation of the PlGF receptor neuropilin 1 to promote cell survival. Since PlGF is expressed in the majority of medulloblastomas, it will be important to further delineate the effect of a Smo blockade in the development of such tumors [111].

4.2 *Smo, Cilia, and Medulloblastoma*

Hh-dependent tumorigenesis can be both mediated or suppressed by the primary cilia, depending on the underlying oncogenic events involving active forms of Smo (SmoM2) or Gli2 (Gli2ΔN) [112, 113]. The constitutive localization of SmoM2 to the cilia leads to Gli2 activation and the inhibition of Gli3 repressor formation, which account for the tumorigenic process. Therefore, the absence of cilia impedes SmoM2-mediated activation of the downstream pathway. In contrast, the constitutive activity of Gli2ΔN is independent of primary cilia. Nevertheless, when the cilia are present, mice expressing Gli2ΔN mutant do not develop medulloblastoma possibly due to the presence of Gli3 repressor. In the absence of cilia, Gli3 repressor is not generated and consequently Gli2ΔN induces medulloblastoma. The presence of primary cilia in specific variants of human medulloblastomas is also important from a therapeutic viewpoint: those with high Hh signaling might be responsive to treatments that target the primary cilium [112]. Smo activation and its regulation both by Hh signaling and small molecules, at the level of the primary cilium, were proposed as key steps in tumorigenesis-dependent processes, but are still not sufficiently studied [101].

Recent studies investigated the link between the dysfunction of primary cilia and Hh signaling in human ciliopathies and in related mouse models [20]. Joubert syndrome (JS) and Meckel syndrome (MKS) are ciliopathies displaying severe brain defects including in the cerebellar vermis. In the cerebellum of these patients, GCPs are severely impaired and presumably linked to an abnormal Shh response [114].

4.3 Blocking Smo for Treating Medulloblastomas

Cyclopamine and several other small molecule inhibitors of the Hh pathway were developed and proposed for the treatment of cancers associated with Hh signaling dysfunctions. Most of these molecules, but not all, target Smo [14, 101]. Among them, Cur61414 or HhAntag691 were shown to induce remission in animal models of medulloblastoma. Clinical trials were recently undertaken for the Smo inhibitor GDC-0449, to treat basal cell carcinoma and medulloblastoma in humans [101]. However, a patient affected with a metastatic medulloblastoma refractory to multiple therapies and treated with this molecule, showed resistance to the treatment due to an acquired Smo mutation disrupting the ability of GDC-0449 to bind Smo [115]. A Smo mutation occurring at a homologous position in mouse Smo was also observed in a GDC-0449-resistant mouse model of medulloblastoma [116]. Resistance also developed in mice with medulloblastoma treated with the Smo inhibitor LDE-225, which is also used in clinical trials [117]. These resistances were associated with the amplification of the transcription factor Gli2, linked to Hh signaling, to activate Smo mutations and possibly to the upregulation of the phosphatidylinositol 3-kinase signaling pathway.

4.4 Hh Inhibitors Regulate Smo Trafficking in the Primary Cilia

Smo activation is proposed to first require translocation into the primary cilium followed by a subsequent activation step. Consequently, Smo inhibitors are classified into at least two families of compounds: SANT1-like molecules that block Smo trafficking to the primary cilium and cyclopamine-like molecules that affect Smo activation within the cilium [92]. Interestingly, the mutant protein SmoM2 is less sensitive to cyclopamine than wild-type Smo, whereas SANT1 inhibits both proteins with similar potency [118]. It is anticipated that SmoM2-like mutations, which might arise from drug treatment in the clinic, will therefore be more sensitive to the SANT1-like family of molecules than to the cyclopamine-like family of molecules [101]. Further studies should clarify the molecular and biochemical mechanisms underlying the resistance to Smo inhibitors, particularly at the level of the primary cilium in normal and cancer cells.

4.5 Smo and Cerebellar Progenitor Cell Maintenance

Impaired neurogenesis during brain development is linked to intellectual disability in Down syndrome (DS). Ts65Dn mice are trisomic for contiguous segments of mouse chromosome 16, which is highly homologous to the long arm of human chromosome 21 implicated in DS [119]. In this widely used mouse model of DS, treating newly born mice with the Smo agonist SAG induces a burst of proliferation of GCPs, which are deficient in these animals, and ameliorates cerebellar morphology in the adult [120]. Furthermore, a corresponding improvement in learning and memory tasks associated with cerebellar functions was seen in these mice, demonstrating a causal relationship with brain development. SAG presumably also acts on hippocampal cells, because *N*-methyl-D-aspartate (NMDA) receptor-dependent synaptic plasticity in the adult was partially rescued [120]. However, the impairment of neuronal precursors in the Ts65Dn mouse model is also linked to the inhibitory effect of the Ptc receptor, which would play a key role in regulating the expression of amyloid precursor protein (APP) [121, 122]. These studies provide important insights for understanding DS defects, but are far from suggesting a possible cure: first, Hh signaling in the human brain is not completely understood, and second, Smo agonists for human clinical use are not yet developed.

5 Potential of Smo Agonists in Central Nervous System Diseases

5.1 *Smo and Parkinson's Disease*

Activation of the brain-specific Shh signaling pathway is a potential therapeutic approach for Parkinson's disease, via Hh signaling, against the loss of dopaminergic neurons in the substantia nigra [8]. The injection of the active fragment of Shh in rodent or primate models of Parkinson's disease increases tyrosine hydroxylase immunoreactive neurons in the substantia nigra, preserves dopaminergic axons in the striatum, and is associated with locomotor function improvement [123, 124]. Recently, transgenic mice with a Shh signaling defect in dopaminergic neurons showed progressive degeneration of dopaminergic, cholinergic, and GABAergic neurons of the mesostriatal circuit in adults. Moreover, these mice displayed symptoms of Parkinson's disease as evidenced by cholinergic and dopaminergic neurotransmission imbalance and motor skill deficits. Further pharmacological manipulation of Hh signaling, by injecting the Smo modulators cyclopamine and SAG, led to cholinergic tone modulation, expression of glial cell line-derived neurotrophic factor (GDNF), and the maintenance of cholinergic and fast-spiking GABAergic neurons in the striatum. These data pave the way for investigating the treatment potential of Smo modulators for palliating some physiological defects of Parkinson's disease [125].

5.2 *Nerve Repair, Demyelinating Diseases, and Brain Injuries*

5.2.1 Reactive Astrocytes

Smo regulation could be also involved in reactive gliosis that occurs upon diverse types of injury involving the activation of microglia and other types of glia cells [126]. Interestingly, besides adult NSCs that have structural characteristics of astrocytes, Shh signaling is active in select populations of mature parenchymal astrocytes, and deletion of Smo in astrocytes results in astrogliosis [127]. It has been proposed that once astrocytes are activated next to the injury site, they acquire stem cell properties upon Shh signaling and start to proliferate upon addition of a Smo agonist [128]. Reactive astrocytes have been also shown to give rise to neurons upon brain injury [129]. Thus, further work using Smo-positive and Smo-negative modulators is needed to understand and delineate the therapeutic interest of Smo manipulation upon brain injury.

5.2.2 Oligodendrocyte Progenitors

Oligodendrocytes are responsible for the formation of the myelin sheath, which is needed for fast signal transmission and for neuronal survival. Demyelination occurs in multiple brain diseases, such as multiple sclerosis, but also upon trauma, ischemia, viral infections, and neurodegenerative diseases [130]. After oligodendrocyte death, spontaneous remyelination occurs through the generation of oligodendrocytes originating from two main sources: oligodendrocyte precursor cells (OPCs) from the parenchyme [131] and SVZ-derived progenitors (Fig. 3b) [56, 132]. Thus, in response to demyelination, OPCs are recruited to the lesion site, migrate, proliferate, and differentiate.

During development, Shh is required for the specification of a first wave of OPCs originating in the ventral regions of the spinal cord and the forebrain. In the adult, brain delivery of Shh is followed by an increase in the number of OPCs, and more differentiated cells expressing the neural/glia antigen 2 (NG2) and the proteolipid protein isoform DM20, both in the cerebral cortex and the corpus callosum [133]. This suggests that the pathway modulates the OPC lineage. However, activating Smo with the agonist SAG by direct injection in the corpus callosum of adult mice results in an increase of NG2-positive cells. This effect on cell proliferation may not involve Gli1 but alternatively a noncanonical Hh signaling pathway [134].

The broad activation of the Shh signaling pathway in a mouse model of focal demyelination is also in agreement with such data [56]. In this model, lysophosphatidyl choline (also called lysolecithin or LPC) injection in the corpus callosum results in rapid degeneration (3–7 days) of most oligodendrocytes except the OPCs. Remyelination then occurs and completes after 15–20 days. The delivery of Shh through adenovirus-mediated transfer into the lesioned brain is followed by an increase of OPCs and mature myelinating OL numbers due to survival, proliferation, and differentiation activities. Interestingly, both astrogliosis and macrophage infiltration are reduced. Blocking the Shh pathway using its physiological antagonist HIP results in a decrease of OPC proliferation and differentiation, preventing further repair. Thus, these data argue for a positive role of Shh signaling during demyelination. However, further work is needed to understand the complex roles exerted by Smo in the lesion, since this receptor is upregulated by microglia cells early after the lesion formation, then by OPCs, and to a lesser extent by astrocytes during remyelination [56]. *Smo* transcripts are also upregulated in adult rat facial motor neurons following facial nerve axotomy [135]. Thus, altogether, these data pave the way to further explore the therapeutic potential of Smo modulation in nerve repair or myelin disease treatment. Moreover, such investigations are also necessary to understand if Smo activation protects glucocorticoid-induced brain injury in neonatal mice. This is important for neonates receiving glucocorticoids to treat airway and pulmonary conditions, or during cardiac bypass [136].

5.3 Summary and Future Prospects

The functional roles of Smo in the developing and adult CNS range from regulation of stem and precursor cells in multiple stem cell niches to the modulation of both astrocytic and neuronal activities [137, 138]. Increasing evidence shows that Smo functions depend on the primary cilium. Progress is still to be made for further understanding the molecular and biochemical events controlling Smo trafficking at this important signaling center. Besides genetic studies aimed at characterizing Smo defects in both astrocytes and neurons, pharmacological approaches using small molecule inhibitors or activators of Smo will unravel novel properties of this atypical receptor. Such studies are likely to provide important data, which would further promote the role of Smo in the tissue repair process in brain diseases [133]. In this context, the development of drug-like Smo agonists and antagonists and the understanding of their mechanisms of action, including at the primary cilium, are of great interest not only for treating brain tumors but also for developing novel therapies for brain diseases.

References

1. Dessaud E, McMahon AP, Briscoe J (2008) Pattern formation in the vertebrate neural tube: a Sonic Hedgehog morphogen-regulated transcriptional network. Development 135:2489–2503
2. Varjosalo M, Taipale J (2008) Hedgehog: functions and mechanisms. Genes Dev 22:2454–2472
3. Briscoe J, Therond PP (2013) The mechanisms of Hedgehog signalling and its roles in development and disease. Nat Rev 14:416–429
4. Coni S, Infante P, Gulino A (2013) Control of stem cells and cancer stem cells by Hedgehog signaling: pharmacologic clues from pathway dissection. Biochem Pharmacol 85:623–628
5. Ruat M, Roudaut H, Ferent J, Traiffort E (2012) Hedgehog trafficking, cilia and brain functions. Differentiation 83:S97–S104
6. Charytoniuk D, Porcel B, Rodriguez Gomez J, Faure H, Ruat M, Traiffort E (2002) Sonic Hedgehog signalling in the developing and adult brain. J Physiol Paris 96:9–16
7. Borzillo GV, Lippa B (2005) The Hedgehog signaling pathway as a target for anticancer drug discovery. Curr Top Med Chem 5:147–157
8. Traiffort E, Angot E, Ruat M (2010) Sonic Hedgehog signaling in the mammalian brain. J Neurochem 113:576–590
9. Alvarez-Buylla A, Ihrie RA (2014) Sonic Hedgehog signaling in the postnatal brain. Semin Cell Dev Biol 33:105–111
10. Han YG, Alvarez-Buylla A (2010) Role of primary cilia in brain development and cancer. Curr Opin Neurobiol 20:58–67
11. Suh H, Deng W, Gage FH (2009) Signaling in adult neurogenesis. Annu Rev Cell Dev Biol 25:253–275
12. Ng JM, Curran T (2011) The Hedgehog's tale: developing strategies for targeting cancer. Nat Rev Cancer 11:493–501
13. Low JA, de Sauvage FJ (2010) Clinical experience with Hedgehog pathway inhibitors. J Clin Oncol 28:5321–5326

14. Heretsch P, Tzagkaroulaki L, Giannis A (2010) Modulators of the Hedgehog signaling pathway. Bioorg Med Chem 18:6613–6624
15. Mas C, Ruiz i Altaba A (2010) Small molecule modulation of HH-GLI signaling: current leads, trials and tribulations. Biochem Pharmacol 80:712–723
16. Hadden MK (2013) Hedgehog pathway inhibitors: a patent review (2009–present). Expert Opin Ther Pat 23:345–361
17. Huangfu D, Liu A, Rakeman AS, Murcia NS, Niswander L, Anderson KV (2003) Hedgehog signalling in the mouse requires intraflagellar transport proteins. Nature 426:83–87
18. Goetz SC, Anderson KV (2010) The primary cilium: a signalling centre during vertebrate development. Nat Rev Genet 11:331–344
19. Louvi A, Grove EA (2011) Cilia in the CNS: the quiet organelle claims center stage. Neuron 69:1046–1060
20. Simpson F, Kerr MC, Wicking C (2009) Trafficking, development and Hedgehog. Mech Dev 126:279–288
21. Pepinsky RB, Zeng C, Wen D, Rayhorn P, Baker DP, Williams KP, Bixler SA, Ambrose CM, Garber EA, Miatkowski K, Taylor FR, Wang EA, Galdes A (1998) Identification of a palmitic acid-modified form of human Sonic Hedgehog. J Biol Chem 273:14037–14045
22. Traiffort E, Moya KL, Faure H, Hassig R, Ruat M (2001) High expression and anterograde axonal transport of aminoterminal Sonic Hedgehog in the adult hamster brain. Eur J Neurosci 14:839–850
23. Coulombe J, Traiffort E, Loulier K, Faure H, Ruat M (2004) Hedgehog interacting protein in the mature brain: membrane-associated and soluble forms. Mol Cell Neurosci 25:323–333
24. Mukhopadhyay S, Wen X, Ratti N, Loktev A, Rangell L, Scales SJ, Jackson PK (2013) The ciliary G-protein-coupled receptor Gpr161 negatively regulates the Sonic Hedgehog pathway via cAMP signaling. Cell 152:210–223
25. Kovacs JJ, Whalen EJ, Liu R, Xiao K, Kim J, Chen M, Wang J, Chen W, Lefkowitz RJ (2008) Beta-arrestin-mediated localization of smoothened to the primary cilium. Science 320:1777–1781
26. Dorn KV, Hughes CE, Rohatgi R (2012) A Smoothened-Evc2 complex transduces the Hedgehog signal at primary cilia. Dev Cell 23:823–835
27. Milenkovic L, Scott MP, Rohatgi R (2009) Lateral transport of Smoothened from the plasma membrane to the membrane of the cilium. J Cell Biol 187:365–374
28. Ruiz i Altaba A, Mas C, Stecca B (2007) The Gli code: an information nexus regulating cell fate, stemness and cancer. Trends Cell Biol 17:438–447
29. Riobo NA, Manning DR (2007) Pathways of signal transduction employed by vertebrate Hedgehogs. Biochem J 403:369–379
30. Allen BL, Song JY, Izzi L, Althaus IW, Kang JS, Charron F, Krauss RS, McMahon AP (2011) Overlapping roles and collective requirement for the coreceptors GAS1, CDO, and BOC in SHH pathway function. Dev Cell 20:775–787
31. Izzi L, Levesque M, Morin S, Laniel D, Wilkes BC, Mille F, Krauss RS, McMahon AP, Allen BL, Charron F (2011) Boc and Gas1 each form distinct Shh receptor complexes with Ptch1 and are required for Shh-mediated cell proliferation. Dev Cell 20:788–801
32. Yam PT, Langlois SD, Morin S, Charron F (2009) Sonic Hedgehog guides axons through a noncanonical, Src-family-kinase-dependent signaling pathway. Neuron 62:349–362
33. Belgacem YH, Borodinsky LN (2011) Sonic Hedgehog signaling is decoded by calcium spike activity in the developing spinal cord. Proc Natl Acad Sci U S A 108:4482–4487
34. Teperino R, Amann S, Bayer M, McGee SL, Loipetzberger A, Connor T, Jaeger C, Kammerer B, Winter L, Wiche G, Dalgaard K, Selvaraj M, Gaster M, Lee-Young RS, Febbraio MA, Knauf C, Cani PD, Aberger F, Penninger JM, Pospisilik JA, Esterbauer H (2012) Hedgehog partial agonism drives Warburg-like metabolism in muscle and brown fat. Cell 151:414–426

35. Polizio AH, Chinchilla P, Chen X, Kim S, Manning DR, Riobo NA (2011) Heterotrimeric Gi proteins link Hedgehog signaling to activation of Rho small GTPases to promote fibroblast migration. J Biol Chem 286:19589–19596
36. Jenkins D (2009) Hedgehog signalling: emerging evidence for non-canonical pathways. Cell Signal 21:1023–1034
37. Javelaud D, Pierrat MJ, Mauviel A (2012) Crosstalk between TGF-beta and Hedgehog signaling in cancer. FEBS Lett 586:2016–2025
38. Yam PT, Charron F (2013) Signaling mechanisms of non-conventional axon guidance cues: the Shh, BMP and Wnt morphogens. Curr Opin Neurobiol 23:965–973
39. Brennan D, Chen X, Cheng L, Mahoney M, Riobo NA (2012) Noncanonical Hedgehog signaling. Vitam Horm 88:55–72
40. Bettencourt-Dias M, Hildebrandt F, Pellman D, Woods G, Godinho SA (2011) Centrosomes and cilia in human disease. Trends Genet 27:307–315
41. Rosenbaum JL, Witman GB (2002) Intraflagellar transport. Nat Rev 3:813–825
42. Taschner M, Bhogaraju S, Lorentzen E (2011) Architecture and function of IFT complex proteins in ciliogenesis. Differentiation 83:S12–S22
43. Corbit KC, Aanstad P, Singla V, Norman AR, Stainier DY, Reiter JF (2005) Vertebrate smoothened functions at the primary cilium. Nature 437:1018–1021
44. Haycraft CJ, Banizs B, Aydin-Son Y, Zhang Q, Michaud EJ, Yoder BK (2005) Gli2 and Gli3 localize to cilia and require the intraflagellar transport protein polaris for processing and function. PLoS Genet 1:e53
45. Huangfu D, Anderson KV (2006) Signaling from Smo to Ci/Gli: conservation and divergence of Hedgehog pathways from Drosophila to vertebrates. Development 13(3):3–14
46. Rohatgi R, Milenkovic L, Scott MP (2007) Patched1 regulates Hedgehog signaling at the primary cilium. Science 317:372–376
47. Chamberlain CE, Jeong J, Guo C, Allen BL, McMahon AP (2008) Notochord-derived Shh concentrates in close association with the apically positioned basal body in neural target cells and forms a dynamic gradient during neural patterning. Development 135:1097–1106
48. Friedland-Little JM, Hoffmann AD, Ocbina PJ, Peterson MA, Bosman JD, Chen Y, Cheng SY, Anderson KV, Moskowitz IP (2011) A novel murine allele of intraflagellar transport protein 172 causes a syndrome including VACTERL-like features with hydrocephalus. Hum Mol Genet 20:3725–3737
49. Ayers KL, Therond PP (2010) Evaluating smoothened as a G-protein-coupled receptor for Hedgehog signalling. Trends Cell Biol 20:287–298
50. Ingham PW, Nakano Y, Seger C (2011) Mechanisms and functions of Hedgehog signalling across the metazoa. Nat Rev Genet 12:393–406
51. Endoh-Yamagami S, Evangelista M, Wilson D, Wen X, Theunissen JW, Phamluong K, Davis M, Scales SJ, Solloway MJ, de Sauvage FJ, Peterson AS (2009) The mammalian Cos2 homolog Kif7 plays an essential role in modulating Hh signal transduction during development. Curr Biol 19:1320–1326
52. Liem KF Jr, He M, Ocbina PJ, Anderson KV (2009) Mouse Kif7/Costal2 is a cilia-associated protein that regulates Sonic Hedgehog signaling. Proc Natl Acad Sci U S A 106:13377–13382
53. Kim J, Kato M, Beachy PA (2009) Gli2 trafficking links Hedgehog-dependent activation of smoothened in the primary cilium to transcriptional activation in the nucleus. Proc Natl Acad Sci U S A 106(51):21666–21671
54. Doetsch F, Caille I, Lim DA, Garcia-Verdugo JM, Alvarez-Buylla A (1999) Subventricular zone astrocytes are neural stem cells in the adult mammalian brain. Cell 97:703–716
55. El Waly B, Macchi M, Cayre M, Durbec P (2014) Oligodendrogenesis in the normal and pathological central nervous system. Front Neurosci 8:145
56. Ferent J, Zimmer C, Durbec P, Ruat M, Traiffort E (2013) Sonic Hedgehog signaling is a positive oligodendrocyte regulator during demyelination. J Neurosci 33:1759–1772

57. Charytoniuk D, Traiffort E, Hantraye P, Hermel JM, Galdes A, Ruat M (2002) Intrastriatal Sonic Hedgehog injection increases patched transcript levels in the adult rat subventricular zone. Eur J Neurosci 16:2351–2357
58. Ahn S, Joyner AL (2005) In vivo analysis of quiescent adult neural stem cells responding to Sonic Hedgehog. Nature 437:894–897
59. Ferent J, Cochard L, Faure H, Taddei M, Hahn H, Ruat M, Traiffort E (2014) Genetic activation of Hedgehog signaling unbalances the rate of neural stem cell renewal by increasing symmetric divisions. Stem Cell Rep 3:312–323
60. Angot E, Loulier K, Nguyen-Ba-Charvet KT, Gadeau AP, Ruat M, Traiffort E (2008) Chemoattractive activity of Sonic Hedgehog in the adult subventricular zone modulates the number of neural precursors reaching the olfactory bulb. Stem Cells 26:2311–2320
61. Aimone JB, Li Y, Lee SW, Clemenson GD, Deng W, Gage FH (2014) Regulation and function of adult neurogenesis: from genes to cognition. Physiol Rev 94:991–1026
62. Christian KM, Song H, Ming GL (2014) Functions and dysfunctions of adult hippocampal neurogenesis. Annu Rev Neurosci 37:243–262
63. Han YG, Spassky N, Romaguera-Ros M, Garcia-Verdugo JM, Aguilar A, Schneider-Maunoury S, Alvarez-Buylla A (2008) Hedgehog signaling and primary cilia are required for the formation of adult neural stem cells. Nat Neurosci 11:277–284
64. Breunig JJ, Sarkisian MR, Arellano JI, Morozov YM, Ayoub AE, Sojitra S, Wang B, Flavell RA, Rakic P, Town T (2008) Primary cilia regulate hippocampal neurogenesis by mediating Sonic Hedgehog signaling. Proc Natl Acad Sci U S A 105:13127–13132
65. Lai K, Kaspar BK, Gage FH, Schaffer DV (2003) Sonic Hedgehog regulates adult neural progenitor proliferation in vitro and in vivo. Nat Neurosci 6:21–27
66. Kriegstein A, Alvarez-Buylla A (2009) The glial nature of embryonic and adult neural stem cells. Annu Rev Neurosci 32:149–184
67. Zhao C, Deng W, Gage FH (2008) Mechanisms and functional implications of adult neurogenesis. Cell 132:645–660
68. Silva-Vargas V, Crouch EE, Doetsch F (2013) Adult neural stem cells and their niche: a dynamic duo during homeostasis, regeneration, and aging. Curr Opin Neurobiol 23:935–942
69. Pereira Dias G, Hollywood R, Bevilaqua MC, da Luz AC, Hindges R, Nardi AE, Thuret S (2014) Consequences of cancer treatments on adult hippocampal neurogenesis: implications for cognitive function and depressive symptoms. Neuro Oncol 16:476–492
70. Traiffort E, Charytoniuk D, Watroba L, Faure H, Sales N, Ruat M (1999) Discrete localizations of Hedgehog signalling components in the developing and adult rat nervous system. Eur J Neurosci 11:3199–3214
71. Traiffort E, Charytoniuk DA, Faure H, Ruat M (1998) Regional distribution of Sonic Hedgehog, patched, and smoothened mRNA in the adult rat brain. J Neurochem 70:1327–1330
72. Charytoniuk DA, Traiffort E, Faure H, Salès N, Ruat M (1998) Localisation of cells expressing patched and smoothened mRNA in adult mouse and rat brain. Eur J Neurosci 10:16738
73. Ihrie RA, Shah JK, Harwell CC, Levine JH, Guinto CD, Lezameta M, Kriegstein AR, Alvarez-Buylla A (2011) Persistent Sonic Hedgehog signaling in adult brain determines neural stem cell positional identity. Neuron 71:250–262
74. Petrova R, Garcia AD, Joyner AL (2013) Titration of GLI3 repressor activity by Sonic Hedgehog signaling is critical for maintaining multiple adult neural stem cell and astrocyte functions. J Neurosci 33:17490–17505
75. Roudaut H, Traiffort E, Gorojankina T, Vincent L, Faure H, Schoenfelder A, Mann A, Manetti F, Solinas A, Taddei M, Ruat M (2011) Identification and mechanism of action of the acylguanidine MRT-83, a novel potent Smoothened antagonist. Mol Pharmacol 79:453–460
76. Machold R, Hayashi S, Rutlin M, Muzumdar MD, Nery S, Corbin JG, Gritli-Linde A, Dellovade T, Porter JA, Rubin LL, Dudek H, McMahon AP, Fishell G (2003) Sonic

Hedgehog is required for progenitor cell maintenance in telencephalic stem cell niches. Neuron 39:937–950
77. Palma V, Lim DA, Dahmane N, Sanchez P, Brionne TC, Herzberg CD, Gitton Y, Carleton A, Alvarez-Buylla A, Ruiz i Altaba A (2005) Sonic Hedgehog controls stem cell behavior in the postnatal and adult brain. Development 132:335–344
78. Banerjee SB, Rajendran R, Dias BG, Ladiwala U, Tole S, Vaidya VA (2005) Recruitment of the Sonic Hedgehog signalling cascade in electroconvulsive seizure-mediated regulation of adult rat hippocampal neurogenesis. Eur J Neurosci 22:1570–1580
79. Balordi F, Fishell G (2007) Hedgehog signaling in the subventricular zone is required for both the maintenance of stem cells and the migration of newborn neurons. J Neurosci 27:5936–5947
80. Balordi F, Fishell G (2007) Mosaic removal of Hedgehog signaling in the adult SVZ reveals that the residual wild-type stem cells have a limited capacity for self-renewal. J Neurosci 27:14248–14259
81. Merkle FT, Fuentealba LC, Sanders TA, Magno L, Kessaris N, Alvarez-Buylla A (2014) Adult neural stem cells in distinct microdomains generate previously unknown interneuron types. Nat Neurosci 17:207–214
82. Brill MS, Ninkovic J, Winpenny E, Hodge RD, Ozen I, Yang R, Lepier A, Gascon S, Erdelyi F, Szabo G, Parras C, Guillemot F, Frotscher M, Berninger B, Hevner RF, Raineteau O, Gotz M (2009) Adult generation of glutamatergic olfactory bulb interneurons. Nat Neurosci 12:1524–1533
83. Lledo PM, Alonso M, Grubb MS (2006) Adult neurogenesis and functional plasticity in neuronal circuits. Nat Rev Neurosci 7:179–193
84. Myers BR, Sever N, Chong YC, Kim J, Belani JD, Rychnovsky S, Bazan JF, Beachy PA (2013) Hedgehog pathway modulation by multiple lipid binding sites on the smoothened effector of signal response. Dev Cell 26:346–357
85. Nachtergaele S, Mydock LK, Krishnan K, Rammohan J, Schlesinger PH, Covey DF, Rohatgi R (2012) Oxysterols are allosteric activators of the oncoprotein Smoothened. Nat Chem Biol 8:211–220
86. Nachtergaele S, Whalen DM, Mydock LK, Zhao Z, Malinauskas T, Krishnan K, Ingham PW, Covey DF, Siebold C, Rohatgi R (2013) Structure and function of the smoothened extracellular domain in vertebrate Hedgehog signaling. Elife 2:e01340
87. Nedelcu D, Liu J, Xu Y, Jao C, Salic A (2013) Oxysterol binding to the extracellular domain of smoothened in Hedgehog signaling. Nat Chem Biol 9:557–564
88. Wang J, Lu J, Bond MC, Chen M, Ren XR, Lyerly HK, Barak LS, Chen W (2010) Identification of select glucocorticoids as smoothened agonists: potential utility for regenerative medicine. Proc Natl Acad Sci U S A 107:9323–9328
89. Favaro R, Valotta M, Ferri AL, Latorre E, Mariani J, Giachino C, Lancini C, Tosetti V, Ottolenghi S, Taylor V, Nicolis SK (2009) Hippocampal development and neural stem cell maintenance require Sox2-dependent regulation of Shh. Nat Neurosci 12:1248–1256
90. Li G, Fang L, Fernandez G, Pleasure SJ (2013) The ventral hippocampus is the embryonic origin for adult neural stem cells in the dentate gyrus. Neuron 78:658–672
91. Takeda S, Yonekawa Y, Tanaka Y, Okada Y, Nonaka S, Hirokawa N (1999) Left-right asymmetry and kinesin superfamily protein KIF3A: new insights in determination of laterality and mesoderm induction by kif3A−/− mice analysis. J Cell Biol 145:825–836
92. Rohatgi R, Milenkovic L, Corcoran RB, Scott MP (2009) Hedgehog signal transduction by smoothened: pharmacologic evidence for a 2-step activation process. Proc Natl Acad Sci U S A 106:3196–3201
93. Amador-Arjona A, Elliott J, Miller A, Ginbey A, Pazour GJ, Enikolopov G, Roberts AJ, Terskikh AV (2011) Primary cilia regulate proliferation of amplifying progenitors in adult hippocampus: implications for learning and memory. J Neurosci 31:9933–9944

94. Town T, Breunig JJ, Sarkisian MR, Spilianakis C, Ayoub AE, Liu X, Ferrandino AF, Gallagher AR, Li MO, Rakic P, Flavell RA (2008) The stumpy gene is required for mammalian ciliogenesis. Proc Natl Acad Sci U S A 105:2853–2858
95. Galvin KE, Ye H, Erstad DJ, Feddersen R, Wetmore C (2008) Gli1 induces G2/M arrest and apoptosis in hippocampal but not tumor-derived neural stem cells. Stem Cells 26:1027–1036
96. Bilkei-Gorzo A (2014) Genetic mouse models of brain ageing and Alzheimer's disease. Pharmacol Ther 142:244–257
97. He P, Staufenbiel M, Li R, Shen Y (2014) Deficiency of Patched 1-induced Gli1 signal transduction results in astrogenesis in Swedish mutated APP transgenic mice. Hum Mol Genet 23(24):6512–6527
98. Masdeu C, Bernard V, Faure H, Traiffort E, Ruat M (2007) Distribution of Smoothened at hippocampal mossy fiber synapses. Neuroreport 18:395–399
99. Petralia RS, Schwartz CM, Wang YX, Mattson MP, Yao PJ (2011) Subcellular localization of patched and smoothened, the receptors for Sonic Hedgehog signaling, in the hippocampal neuron. J Comp Neurol 519:3684–3699
100. Jones DT, Jager N, Kool M, Zichner T, Hutter B et al (2012) Dissecting the genomic complexity underlying medulloblastoma. Nature 488:100–105
101. Scales SJ, de Sauvage FJ (2009) Mechanisms of Hedgehog pathway activation in cancer and implications for therapy. Trends Pharmacol Sci 30:303–312
102. Pritchard JI, Olson JM (2008) Methylation of PTCH1, the Patched-1 gene, in a panel of primary medulloblastomas. Cancer Genet Cytogenet 180:47–50
103. Corcoran RB, Bachar Raveh T, Barakat MT, Lee EY, Scott MP (2008) Insulin-like growth factor 2 is required for progression to advanced medulloblastoma in patched1 heterozygous mice. Cancer Res 68:8788–8795
104. Parathath SR, Mainwaring LA, Fernandez LA, Campbell DO, Kenney AM (2008) Insulin receptor substrate 1 is an effector of Sonic Hedgehog mitogenic signaling in cerebellar neural precursors. Development 135:3291–3300
105. Yang ZJ, Ellis T, Markant SL, Read TA, Kessler JD, Bourboulas M, Schuller U, Machold R, Fishell G, Rowitch DH, Wainwright BJ, Wechsler-Reya RJ (2008) Medulloblastoma can be initiated by deletion of patched in lineage-restricted progenitors or stem cells. Cancer Cell 14:135–145
106. Li P, Du F, Yuelling LW, Lin T, Muradimova RE, Tricarico R, Wang J, Enikolopov G, Bellacosa A, Wechsler-Reya RJ, Yang ZJ (2013) A population of Nestin-expressing progenitors in the cerebellum exhibits increased tumorigenicity. Nat Neurosci 16:1737–1744
107. Amakye D, Jagani Z, Dorsch M (2013) Unraveling the therapeutic potential of the Hedgehog pathway in cancer. Nat Med 19:1410–1422
108. Dahmane N, Ruiz-i-Altaba A (1999) Sonic Hedgehog regulates the growth and patterning of the cerebellum. Development 126:3089–3100
109. Rowitch DH, S-J B, Lee SM, Flax JD, Snyder EY, McMahon AP (1999) Sonic Hedgehog regulates proliferation and inhibits differentiation of CNS precursor cells. J Neurosci 19:8954–8965
110. Petralia RS, Wang YX, Mattson MP, Yao PJ (2012) Subcellular distribution of patched and smoothened in the cerebellar neurons. Cerebellum 11:972–981
111. Snuderl M, Batista A, Kirkpatrick ND, Ruiz de Almodovar C et al (2013) Targeting placental growth factor/neuropilin 1 pathway inhibits growth and spread of medulloblastoma. Cell 152:1065–1076
112. Han YG, Kim HJ, Dlugosz AA, Ellison DW, Gilbertson RJ, Alvarez-Buylla A (2009) Dual and opposing roles of primary cilia in medulloblastoma development. Nat Med 15:1062–1065
113. Wong SY, Seol AD, So PL, Ermilov AN, Bichakjian CK, Epstein EH Jr, Dlugosz AA, Reiter JF (2009) Primary cilia can both mediate and suppress Hedgehog pathway-dependent tumorigenesis. Nat Med 15:1055–1061

114. Aguilar A, Meunier A, Strehl L, Martinovic J, Bonniere M, Attie-Bitach T, Encha-Razavi F, Spassky N (2012) Analysis of human samples reveals impaired SHH-dependent cerebellar development in Joubert syndrome/Meckel syndrome. Proc Natl Acad Sci U S A 109:16951–16956
115. Rudin CM, Hann CL, Laterra J, Yauch RL, Callahan CA, Fu L, Holcomb T, Stinson J, Gould SE, Coleman B, LoRusso PM, Von Hoff DD, de Sauvage FJ, Low JA (2009) Treatment of medulloblastoma with Hedgehog pathway inhibitor GDC-0449. N Engl J Med 361:1173–1178
116. Yauch RL, Dijkgraaf GJ, Alicke B, Januario T, Ahn CP, Holcomb T, Pujara K, Stinson J, Callahan CA, Tang T, Bazan JF, Kan Z, Seshagiri S, Hann CL, Gould SE, Low JA, Rudin CM, de Sauvage FJ (2009) Smoothened mutation confers resistance to a Hedgehog pathway inhibitor in medulloblastoma. Science 326:572–574
117. Buonamici S, Williams J, Morrissey M, Wang A, Guo R, Vattay A, Hsiao K, Yuan J, Green J, Ospina B, Yu Q, Ostrom L, Fordjour P, Anderson DL, Monahan JE, Kelleher JF, Peukert S, Pan S, Wu X, Maira SM, Garcia-Echeverria C, Briggs KJ, Watkins DN, Yao YM, Lengauer C, Warmuth M, Sellers WR, Dorsch M (2010) Interfering with resistance to smoothened antagonists by inhibition of the PI3K pathway in medulloblastoma. Sci Transl Med 2:51–70
118. Chen JK, Taipale J, Cooper MK, Beachy PA (2002) Inhibition of Hedgehog signaling by direct binding of cyclopamine to smoothened. Genes Dev 16:2743–2748
119. Costa AC, Scott-McKean JJ (2013) Prospects for improving brain function in individuals with down syndrome. CNS Drugs 27:679–702
120. Das I, Park JM, Shin JH, Jeon SK, Lorenzi H, Linden DJ, Worley PF, Reeves RH (2013) Hedgehog agonist therapy corrects structural and cognitive deficits in a down syndrome mouse model. Sci Transl Med 5:201ra120
121. Trazzi S, Fuchs C, Valli E, Perini G, Bartesaghi R, Ciani E (2013) The amyloid precursor protein (APP) triplicated gene impairs neuronal precursor differentiation and neurite development through two different domains in the Ts65Dn mouse model for Down syndrome. J Biol Chem 288:20817–20829
122. Trazzi S, Mitrugno VM, Valli E, Fuchs C, Rizzi S, Guidi S, Perini G, Bartesaghi R, Ciani E (2011) APP-dependent up-regulation of Ptch1 underlies proliferation impairment of neural precursors in Down syndrome. Hum Mol Genet 20:1560–1573
123. Tsuboi K, Shults CW (2002) Intrastriatal injection of Sonic Hedgehog reduces behavioral impairment in a rat model of Parkinson's disease. Exp Neurol 173:95–104
124. Dass B, Iravani MM, Jackson MJ, Engber TM, Galdes A, Jenner P (2002) Behavioural and immunohistochemical changes following supranigral administration of Sonic Hedgehog in 1-methyl-4-phenyl-1,2,3,6-tetrahydropyridine-treated common marmosets. Neuroscience 114:99–109
125. Gonzalez-Reyes LE, Verbitsky M, Blesa J, Jackson-Lewis V, Paredes D, Tillack K, Phani S, Kramer ER, Przedborski S, Kottmann AH (2012) Sonic Hedgehog maintains cellular and neurochemical homeostasis in the adult nigrostriatal circuit. Neuron 75:306–319
126. Dimou L, Gotz M (2014) Glial cells as progenitors and stem cells: new roles in the healthy and diseased brain. Physiol Rev 94:709–737
127. Garcia AD, Petrova R, Eng L, Joyner AL (2010) Sonic Hedgehog regulates discrete populations of astrocytes in the adult mouse forebrain. J Neurosci 30:13597–13608
128. Sirko S, Behrendt G, Johansson PA, Tripathi P, Costa M, Bek S, Heinrich C, Tiedt S, Colak D, Dichgans M, Fischer IR, Plesnila N, Staufenbiel M, Haass C, Snapyan M, Saghatelyan A, Tsai LH, Fischer A, Grobe K, Dimou L, Gotz M (2013) Reactive glia in the injured brain acquire stem cell properties in response to Sonic Hedgehog [corrected]. Cell Stem Cell 12:426–439
129. Guo Z, Zhang L, Wu Z, Chen Y, Wang F, Chen G (2013) In vivo direct reprogramming of reactive glial cells into functional neurons after brain injury and in an Alzheimer's disease model. Cell Stem Cell 14:188–202

130. Edgar N, Sibille E (2012) A putative functional role for oligodendrocytes in mood regulation. Transl Psychiatry 2:e109
131. Dawson MR, Polito A, Levine JM, Reynolds R (2003) NG2-expressing glial progenitor cells: an abundant and widespread population of cycling cells in the adult rat CNS. Mol Cell Neurosci 24:476–488
132. Menn B, Garcia-Verdugo JM, Yaschine C, Gonzalez-Perez O, Rowitch D, Alvarez-Buylla A (2006) Origin of oligodendrocytes in the subventricular zone of the adult brain. J Neurosci 26:7907–7918
133. Loulier K, Ruat M, Traiffort E (2006) Increase of proliferating oligodendroglial progenitors in the adult mouse brain upon Sonic Hedgehog delivery in the lateral ventricle. J Neurochem 98:530–542
134. Mierzwa AJ, Sullivan GM, Beer LA, Ahn S, Armstrong RC (2014) Comparison of cortical and white matter traumatic brain injury models reveals differential effects in the subventricular zone and divergent Sonic Hedgehog signaling pathways in neuroblasts and oligodendrocyte progenitors. ASN Neuro. doi:10.1177/1759091414551782
135. Akazawa C, Tsuzuki H, Nakamura Y, Sasaki Y, Ohsaki K, Nakamura S, Arakawa Y, Kohsaka S (2004) The upregulated expression of Sonic Hedgehog in motor neurons after rat facial nerve axotomy. J Neurosci 24:7923–7930
136. Heine VM, Rowitch DH (2009) Hedgehog signaling has a protective effect in glucocorticoid-induced mouse neonatal brain injury through an 11betaHSD2-dependent mechanism. J Clin Invest 119:267–277
137. Bezard E, Baufreton J, Owens G, Crossman AR, Dudek H, Taupignon A, Brotchie JM (2003) Sonic Hedgehog is a neuromodulator in the adult subthalamic nucleus. FASEB J 17:2337–2338
138. Pascual O, Traiffort E, Baker DP, Galdes A, Ruat M, Champagnat J (2005) Sonic Hedgehog signalling in neurons of adult ventrolateral nucleus tractus solitarius. Eur J Neurosci 22:389–396

Top Med Chem (2015) 16: 173–206
DOI: 10.1007/7355_2014_70

Published online: 20 September 2014

The Role of Smoothened and Hh Signaling in Neovascularization

Raffaella Soleti, Ramaroson Andriantsitohaina, and Maria Carmen Martínez

Abstract New vessel formation plays a key role not only in physiological processes such as embryonic development and wound repair but also during several pathological situations. In this respect, favoring neovascularization represents a promising therapeutic approach that would allow inducing tissue repair. Among the candidate proteins able to modulate neovascularization, evidence show that the administration of recombinant hedgehog (Hh) protein, gene, or cell therapy based on Hh transfer or using extracellular vesicles as vectors enhance new vessel formation. Here, we summarized the role of Hh pathway on angiogenesis and its therapeutic potential during myocardial infarction and diabetes.

Keywords Extracellular vesicles, Hedgehog, Neovascularization, Smoothened

Contents

R. Soleti
INSERM UMR1063, Stress oxydant et pathologies métaboliques, Angers, France

R. Andriantsitohaina
INSERM UMR1063, Stress oxydant et pathologies métaboliques, Angers, France

Centre Hospitalier Universitaire d'Angers, Angers, France

M.C. Martínez (✉)
INSERM UMR1063, Stress oxydant et pathologies métaboliques, Angers, France

Institut de Biologie en Santé, 4 rue Larrey, Angers, France
e-mail: carmen.martinez@univ-angers.fr

Abbreviations

AMPK	AMP-activated protein kinase
Ang(1, 2)	Angiopoietin(1, 2)
CECs	Circulating endothelial cells
Dhh	Desert hedgehog
ECM	Extracellular matrix
eNOS	Endothelial nitric oxide synthase
ERK	Extracellular signal-regulated kinase
FAK	Focal adhesion kinase
FGF	Fibroblast growth factor
Gli(1, 2, 3)	Glioma-associated oncogenes(1, 2, 3)
GPCR	G-protein-coupled receptors
Hh	Hedgehog
HIF-1	Hypoxia-inducible factor-1
Ihh	Indian hedgehog
iNOS	Inducible nitric oxide synthase
JNK	c-Jun N-terminal kinase
LGR5	Leucine-rich repeat G-protein-coupled receptor 5
LMPs	Lymphocytic microparticles
MAPK	Mitogen-activated protein kinase
miRs	MicroRNAs
MPs	Microparticles
MPs^{Shh+}	Microparticles expressing sonic hedgehog
mRNA	Messenger RNA
NO	Nitric oxide
PDGF	Platelet-derived growth factor
PECAM-1	Platelet-endothelial cell-adhesion molecule-1
phShh	Plasmid encoding the sonic hedgehog human gene

PI3K	Phosphatidylinositol 3-kinase
PKC	Protein kinase C
Ptc	Patched
ROCK	Rho-associated protein kinase
ROS	Reactive oxygen species
SDF-1	Stromal cell-derived factor-1
Shh	Sonic hedgehog
Smo	Smoothened
TGF-β	Transforming growth factor β
TSP1	Thrombospondin1
TYMP	Thymidine phosphorylase
VE-cadherin	Vascular endothelial cadherin
VEGF	Vascular endothelial growth factor
VEGFR(−1, −2)	Vascular endothelial growth factor receptor(−1, −2)

1 Introduction

In the vertebrate embryo, the first functional organ system developed is the cardiovascular system, which is in charge of maintaining metabolic homeostasis by supplying oxygen and nutrients and removing waste products. During development, blood vessels arise through two processes: (1) vasculogenesis, a de novo process by which progenitor stem cells differentiate and give rise to a vascular network, and (2) angiogenesis, the process by which endothelial cells of mature vascular network proliferate, migrate, and remodel into neovessels that grow into the initially avascular tissue. During the adult life, new vessels develop mainly through angiogenesis, although vasculogenesis also may take place [1]. The genesis of new vessels requires a lot of interactions that must be regulated because of interplay between different cells, growth factors, extracellular matrix (ECM) components, and extracellular vesicles.

Besides, one of the most implicated signals in vasculature development is hedgehog (Hh) pathway [2]. Furthermore, during the adult life, Hh cascade directs the formation or the persistence of certain stem and precursor cell populations [3–5], maintenance of tissue homeostasis, and tissue repair during chronic persistent inflammation [6–8]. In addition, deficits in one or more components of Hh signaling translate into developmental defects [9, 10], and unrepressed signaling underlies several cancers [11–16]. Therefore, the activation of Hh signaling in both embryonic and adult life leads to increased neovascularization [2, 17]. This ability of Hh has been evaluated by observing expression of endogenous proteins, using recombinant proteins, gene transfer, or genetically modified mice. In addition, other studies showed Hh involvement in new vessel formation using extracellular vesicles as vectors.

2 Neovascularization

2.1 General Aspects

The development of different tissues needs vessel genesis for the supply of oxygen and nutrients as well as removal of waste products [18]. At the early study, embryo growth occurs in the absence of vascularization, because passive diffusion is sufficient to allow nutrition [19]. Nevertheless, a rapid evolution of an elaborate network of capillary plexuses and blood vessels occurs.

Blood vessel genesis is a complex biological process, which requires the interaction of multiple pathways. During development, two mechanisms allow blood vessel formation: vasculogenesis and angiogenesis (Fig. 1). From an etymological point of view, these terms are identical; however, from a biological point of view, they describe two different processes.

Vasculogenesis consists of capillary formation from differentiation of mesodermal precursors, called angioblasts, which differentiate into endothelial cells forming a vascular network [20], whereas angiogenesis is the formation of new vessels from existing vessels. During embryonic development, mesodermal stem

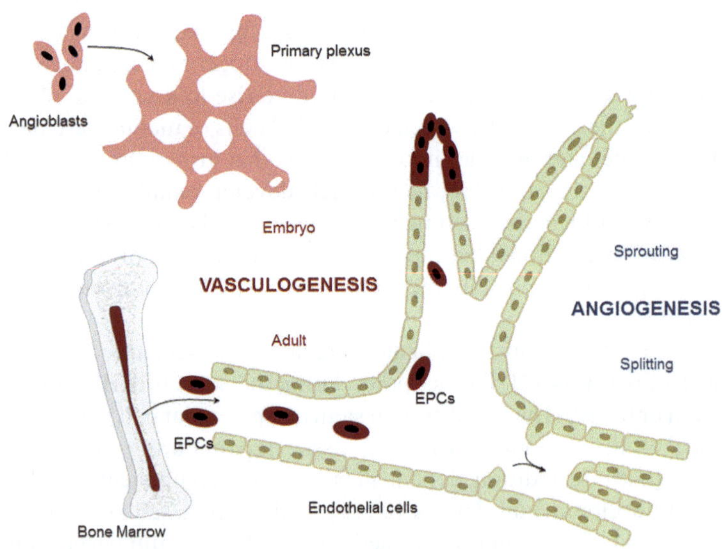

Fig. 1 Processes of neovascularization. Vessel formation occurs to two main mechanisms: vasculogenesis and angiogenesis. The first one takes place in the embryo to allow de novo formation of primary plexus from angioblasts and during adult life contributes to vessel repair upon endothelial precursor cells (EPCs) mobilization from bone marrow. Angiogenesis is defined as vessel growth from a preexisting vessel. Sprouting angiogenesis consists in formation of new branches from preexisting vessels. However, splitting angiogenesis defines the process in which pillars developed within vessels subsequently fuse, delineating new vascular entities or resulting in vessel remodeling

cells differentiate into hemangioblast, which are characterized by the expression of VEGFR-2 (vascular endothelial growth factor receptor-2) and which give rise to hematopoietic lineage and to Notch signaling-specified angioblasts [21]. They migrate from the posterior primitive streak into extraembryonic ectoderm, yolk sac and allantois, and embryonic ectoderm. In the yolk sac, they aggregate to form a primary capillary plexus. Under the activation of vascular endothelial growth factor (VEGF), sonic hedgehog (Shh), and Notch signaling, intraembryonic angioblasts aggregate into the dorsal aorta or cardinal vein and together with the extraembryonic plexus form a de novo vascular network [22]. This novel structure is stabilized by recruitment of other cells, like mural cells or pericytes. Maturation of nascent vessels is mediated by platelet-derived growth factor (PDGF), angiopoietins (Ang1 and 2), and transforming growth factor β (TGF-β) [23].

Angiogenesis is the process of new blood vessel formation from preexisting vessels, which can occur by two principal mechanisms: sprouting and intussusceptions [24, 25].

Vasculogenesis and angiogenesis possess common characteristics; therefore, the distinction between two processes is not categorical. Both need endothelial cell proliferation, migration, and reorganization of newly formed structures and utilize analogous ECM adhesive mechanisms [26]. Moreover, they are not reciprocally exclusive, since angioblasts can be integrated into expanding preexisting blood vessels [27].

During adult life, neovascularization occurs mainly through angiogenesis, although vasculogenesis plays physiological and pathological roles. As a general notion, genesis of the vasculature of most organs occurs by angiogenesis; however, development of the vasculature of certain endodermal organs (liver, lungs, pancreas, stomach/ intestine, and spleen) occurs by vasculogenesis [28]. The existence of a postnatal vasculogenesis is sustained by the observation that both endothelial cells and endothelial progenitor cells coexist in the circulation. Moreover, endothelial progenitor cells are also recruited to sites of neovascularization in mature mammals from a circulating, bone marrow-derived population of progenitor cells [29].

Sprouting angiogenesis consists of highly regulated multi-step process. Hypoxia, inflammation, and mechanical factors are able to initiate new vessel formation. Moreover, they activate endothelial cells, by initiating autocrine or paracrine production and release of growth factors or cytokines [30]. These events cause dissolution of adherent junctions and then increased permeability [31]. Moreover, from the early stage of vessel formation to the later study of capillary elongation, endothelial cells proliferate. Concomitantly, the proteolysis of basement membrane components takes place to allow the invasion and migration of endothelial cells into the surrounding interstitial matrix. Also, endothelial cells line and form the lumen of all blood vessels. This process occurs in a cord of endothelial cells, during both vasculogenesis and angiogenesis. It involves a complex molecular mechanism composed of endothelial cell repulsion at the cell–cell contacts within the endothelial cell cords, junctional rearrangement, and endothelial cell shape changes [32]. The new capillary channel forms an anastomosis with a preexisting capillary, creating a new patent capillary. The final stage requires stabilization of the capillary through the construction of basement membrane, adherent junctions, and cessation of endothelial cell activation [30].

The other type of angiogenesis is called intussusceptive or splitting. It consists of formation of an endothelial lined pillar that extends into the middle of a vessel, followed by the expansion of pillar diameter, resulting in the splitting of the vessel. There are three major forms of intussusceptive angiogenesis. The intussusceptive microvascular growth is the mechanism allowing capillary expansion and shaping vessel morphology. The intussusceptive arborization favors the establishment of vessel hierarchies by the formation and fusion of transcapillary pillars from unremodeled capillaries. Thus, initially large capillaries are parsed into proximal feeding vessels that extend into progressively smaller distal capillaries. In intussusceptive branching remodeling, transcapillary pillars are formed at vessel bifurcations which eventually fuse to the connective tissue to decrease branch angle [33, 34].

Vessel formation is followed by remodeling which is usually a term used to describe a change in the size of a vessel (inward remodeling, which is a reduction in vessel diameter, and outward remodeling, which is an enlargement of vessel diameter) to enable vessels to cope with changes in blood flow [35, 36]. Once vessel formation and remodeling are complete, vessel stability and tissue homeostasis become crucial for a new vessel. This is partially achieved through the recruitment of smooth muscle cells and pericytes collectively known as mural cells to the maturing vessel wall [37]. The interaction between endothelial cells, ECM, and mesenchymal cells is essential to form a stable vasculature. In this stage, communication between different cells is sustained by Ang system [38]. Endothelial cells, through synthesis and secretion of PDGF, a mitogen chemoattractant for a variety of mesenchymal cells, contribute to this phase. Subsequent differentiation of mural precursor cells into pericytes and smooth muscle cells is probably due to a process dependent to cell contact [18]. On endothelial cell–mural cell contact, a latent form of TGF-β, produced by both types of cells, is activated in a plasmin-mediated process and induces changes in myofibroblasts and pericytes. These mechanisms contribute to the formation of a quiescent vessel, ECM production, and maintenance of growth control. Finally, the functional modifications of the largest arteries, such as addiction of a thick muscular coat concomitant with the acquisition of viscoelastic and vasomotor properties, are referred to as arteriogenesis. With the maturation of vascular network, endothelial cells acquire highly specialized characteristics to assure functional needs within specific tissues and organs.

Angiogenesis is a complex process controlled by the balance of counteracting pro- and antiangiogenic factors. When angiogenic activators are more expressed than angiogenic inhibitors, the angiogenic switch will be turned on and angiogenesis will proceed. This situation happens in physiological conditions as in the case of normal wound healing or female endometrial repair. However, in some pathological cases, as in tumor growth, the angiogenic switch remains in the “ON” mode which leads to “nonstop” formation of new blood vessels [39]. Many diseases, including cancer (both solid and hematologic tumors), cardiovascular pathology (atherosclerosis), chronic inflammation (rheumatoid arthritis, Crohn's disease), diabetic retinopathy, psoriasis, and endometriosis [18], are characterized by excessive vascularization. In contrast, other pathologies are associated with impairment of new vasculature formation. Hence, tissue damage after reperfusion of ischemic tissue or cardiac failure or diabetes needs formation of new collateral vessels to improve disease conditions [40, 41].

Under both physiological and pathological conditions, formation of new vessel is governed by growth factors, able to induce, promote, and/or interfere with all phases of angiogenesis [42, 43]. Hypoxia is the most important environmental factor able to initiate neovascularization, by triggering production and activation of growth factors, their receptors, and intracellular signaling mediators.

2.2 *Classical Angiogenic Mediators*

Angiogenesis is orchestrated by many mediators generated from different cells, under both physiological and pathological conditions. These mediators stimulate vessel formation directly by interacting with receptors on endothelial cell surface or indirectly by attracting and activating other cells inducing them to produce angiogenic inducers. Among the numerous angiogenic players, the best known are the family of VEGF [44], hypoxia-inducible factor (HIF-1) [45], and nitric oxide (NO) [46].

The master regulator of angiogenesis and vascular permeability is the VEGF. The downstream signals of VEGF in the vascular endothelium are mediated by two tyrosine kinase receptors, VEGFR-1 and VEGFR-2.

Hypoxia increases VEGF expression via upregulation of HIF-1, as occur in acute human skeletal muscle ischemia [47]. VEGF promotes endothelial cell survival through activation of phosphatidylinositol 3-kinase (PI3K)/Akt pathway and through association with $\alpha_v\beta_3$ integrin and activation of focal adhesion kinase (FAK) [48]. Moreover, VEGF induces endothelial cell proliferation and migration through numerous pathways, including activation of the mitogen-activated protein kinase (MAPK), extracellular signal-regulated kinase (ERK), p38 and c-jun N-terminal kinase (JNK), and Rho GTPase family members [48]. VEGF is one of the strongest known inducers of vascular permeability, which occurs only within minutes after its exposure, probably as the result of efficient simultaneous NO and prostacyclin production [49, 50].

Hypoxia is the well-studied external stimulus activating angiogenesis and represents a key signal for the induction of this process [51]. Hypoxia triggers in affected cells the accumulation of HIF-1, the cellular hypoxia sensor, which is the key element in the process of oxygen homeostasis in vertebrate cells and in the reestablishment of blood vessels in hypoxic areas, such as localized ischemia and tumors [52, 53]. HIF-1 is a master regulator of hypoxic/ischemic vascular responses, driving transcriptional activation of hundreds of genes involved in vascular reactivity, angiogenesis, arteriogenesis, and the mobilization and homing of bone marrow-derived angiogenic cells [45].

NO is a ubiquitous, water-soluble, free radical gas, which, in addition to its well-recognized vasodilatory properties, plays a key role in various physiological and pathological conditions. NO signaling is required for angiogenesis in tumor and wound healing environments, and several angiogenesis factors stimulate this signaling node [54].

2.3 Nonclassical Angiogenic Mediators

In addition to the classic factors, microRNAs (miRs), short noncoding RNAs, and extracellular vesicles (see below) are involved in the modulation of neovascularization by acting on various aspects of the angiogenic process.

miRs are highly conserved, small noncoding RNAs composed of 20–24 nucleotides that control cellular function either by degrading mRNAs or by inhibiting their translation. miRs can be located in the introns of coding or noncoding genes or in exons and are regulated by independent promoters or can be transcribed as a cluster that shares a common promoter. They regulate different aspects of angiogenesis including cell proliferation, matrix remodeling, cell migration, and vessel maturation. Numerous miRs have been shown since then to play a role in angiogenesis by targeting both positive and negative angiogenic regulators, whereas some are expressed by endothelial cells in response to hypoxia or VEGF [55].

2.4 Cells Contributing to Vessel Formation

Endothelial cells are a heterogeneous population, and their growth, differentiation, migration, and survival resulted to be regulated, in a complex manner, by a combination of the surrounding ECM, cell–cell contacts, growth factors, and mechanical cues. As a consequence, the angiogenic process can be influenced and modulated by the interaction of endothelial cells with other cells and with the environment [56]. Moreover, there is considerable phenotypic variation among endothelial cells from different sources, different locations within the same organ, different locations within the same vessel, and different vessel sizes. Therefore, endothelial cells from varying sources exhibit different behaviors in terms of angiogenic potential, as well as molecular permeability, homeostasis, regulation of vascular tone, and even immune tolerance [56].

Elevated levels of circulating endothelial cells (CECs) in the peripheral blood are found in a variety of conditions characterized by vascular injury or vessel formation. In fact, in healthy volunteers, CECs are hardly detectable, whereas the number of CEC correlated with the degree of endothelial injury or neovascularization [57]. Activation of apoptosis-, cytokine-, and protease-mediated injury, defective endothelial cell adhesion, imbalance in pro-angiogenic and antiangiogenic factors, mechanical injury, and drug toxicity represent some mechanisms leading to endothelial cell detachment, and then to their entry in the circulation [58–65]. Therefore, the phenotype and number of CECs may serve as diagnostic or prognostic parameters of vascular injury and tumor growth.

Precursors of endothelial cells, the endothelial progenitor cells, have been identified as participant in angiogenesis. The study of endothelial progenitor cells has received great attention for both their relevance in pathophysiological conditions and their potential as therapeutic agents [66, 67]. Endothelial progenitor cells are a subtype of stem cells with high proliferative potential that are capable of differentiating into mature endothelial cells and contributing to neovascularization. Endothelial progenitor cells are found mainly in the bone marrow in adults, but they are also detected in the peripheral blood; moreover, they are detected in fetal liver and in umbilical cord blood [68].

Circulating endothelial progenitor cells can be subdivided into two main categories: pro-angiogenic hematopoietic cells (early endothelial progenitor cells) and the non-hematopoietic endothelial progenitor cells or outgrowth endothelial cells (late endothelial progenitor cells). While the former originate from bone marrow and represent a pro-vasculogenic subpopulation of hematopoietic stem cells, the latter are not hematopoietic stem-derived cells. These cells can be isolated from blood or tissue samples and amplified via the help of successive culture and distinguished by their obvious endothelial cell-like phenotype or differentiation capability into endothelial cell-like phenotype [69].

Both populations are capable of inducing neovascularization, but the mode of action differs. Whereas early outgrowth endothelial progenitor cells have limited capacity for population doubling and induce only transient angiogenesis, late outgrowth endothelial progenitor cells can expand to more than 100 population doublings. Cell therapy with both populations results in the enhanced engraftment and neovascularization in hindlimb ischemia [70–72]. Moreover, early outgrowth endothelial progenitor cells exert an angiogenic effect mainly by secretor products, whereas late outgrowth cells were thought to produce the effect by direct engraftment [73].

Residing in the bone marrow, endothelial progenitor cells express CD133, CD34, and VEGFR-2 resembling an angioblastic phenotype. After entry into the circulation, endothelial progenitor cells mature and start to express endothelial cell-specific markers such as CD31 (platelet-endothelial cell-adhesion molecule [PECAM-1]), CD146 (vascular endothelial cadherin [VE-cadherin]), von Willebrand factor, and endothelial nitric oxide synthase (eNOS), whereas they gradually lose expression of CD133 and CD34 (except for microvascular endothelial cells that retain CD34 expression). Once endothelial progenitor cells enter the circulation, these cells are referred to as circulating endothelial progenitor cells that are capable of generating outgrowth endothelial cells [74].

Endothelial progenitor cells contribute to re-endothelialization of transplanted tissue. In fact, models have shown that the endothelium of aortic allografts consists entirely of recipient-derived cells [75]. On the contrary, in cardiac transplant models very few endothelial cells of recipient origin have been observed. It is possible that the observed difference is due to the use of immunosuppressants in the cardiac but not aortic graft models [74]. Indeed, the number of circulating endothelial progenitor cells is reduced in cardiac allograft patients with established transplant vasculopathy [76], indicating that sequestration of these cells to the graft site is plausible. Indeed, up to a third of endothelial progenitor cells may be recipient derived [77].

Under steady-state conditions, endothelial progenitor cells are maintained in an undifferentiated and quiescent state but are mobilized following physiological stress and subsequently home to sites of vascular damage [78–80]. Indeed, there is a significant and rapid increase in the number of circulating endothelial progenitor cells following traumatic vascular injury [63]. Vascular damage involves endothelial denudation and platelet accumulation. Platelets, along with neighboring endothelial and smooth muscle cells, secrete various growth factors and cytokines/chemokines, including PDGF, important for smooth muscle cell migration and proliferation, and VEGF and stromal cell-derived factor-1 (SDF-1) which are important for the mobilization of endothelial progenitor cells circulating in the blood and promoting progenitor cell translocation from the quiescent vascular bone marrow niche. Once mobilized, endothelial progenitor cells migrate through the circulation to the site of damage where chemokines and integrins mediate their recruitment and there is evidence for both luminal and adventitial routes of entry [81].

Different studies in both animals and humans demonstrated that endothelial progenitor cells contribute to neovascularization and re-endothelialization [82]. These demonstrations support the possibility that exogenous therapeutic endothelial progenitor cells may provide further benefits to endogenous repair mechanisms, by counteracting ongoing injury and replacing dysfunctional or damaged endothelium.

Therefore, to favor neovascularization is necessary in case pathologies associated with ischemia, on the contrary, target vessel growth essential to treat pathologies characterized by excessive angiogenesis.

2.5 Alterations in Angiogenesis

Angiogenesis is historically linked to cancer, as well as arthritis and psoriasis; however, it occurs in the adult in physiological conditions as during the wound healing, the skeletal growth, or the menstrual cycle. However, it has become evident that perturbations of this process, whether insufficient or excessive vascularization, contribute to the pathogenesis of different disorders such as diabetes, heart and brain ischemia, and cancer. Diabetes mellitus is a chronic metabolic disease characterized by the presence of hyperglycemia, which can lead to many complications. A large number of cellular and subcellular changes on vessels are associated with nephropathy, retinopathy, neuropathy, impaired wound healing, and accelerated atherosclerosis. Angiogenesis plays an ambiguous role in diabetes mellitus [83]. In retinopathy, nephropathy, and atherosclerotic plaque, there is excessive angiogenesis, whereas in wound healing and myocardial perfusion, blood vessel growth is impaired. Excessive angiogenesis leads to the increased risk of cardiovascular events; the growth of vasa vasorum inside vascular wall is stimulated, leading to bleeding, plaque instability, and consequent rupture. In contrast, in diabetic foot disease there is a great reduction in angiogenesis which is at the basis of impaired arteriole growth that leads to the deficit of myocardial perfusion often seen in diabetic patients [84].

Diabetic retinopathy is a clinically well-defined, chronic microvascular complication affecting patients with diabetes, which is characterized by gradually progressive alterations in the retinal microvasculature, leading to areas of retinal nonperfusion, increased vasopermeability, and, in response to retinal nonperfusion, pathologic intraocular proliferation of retinal vessels [85]. Diabetic retinopathy starts with the loss of the two cellular components of retinal capillaries: the pericyte, a vessel support cell, and the endothelial cell. The first pathological change in diabetic retinopathy is decreased pericyte coverage of retinal capillaries and acellular capillaries representing apoptosis of pericytes and endothelial cells. In response to progressive retinal capillary dropout, the ischemic retina mounts an angiogenic response from the surrounding capillaries leading to proliferative diabetic retinopathy [86]. Impaired perfusion and retinal ischemia then causes upregulation of angiogenic molecules including VEGF, erythropoietin, and other vascular growth factors. These factors promote proliferative diabetic retinopathy and lead to increased vascular leakage [87]. One of the well-established clinical advances in preventing vascular complications of diabetes includes agents targeting VEGF which decreases the progression of diabetic retinopathy [88].

On the other hand, impaired wound healing represents another complication of diabetes mellitus. The mechanisms associated concern the enhancement of reactive oxygen species (ROS), the decrease of NO production, the deficit of chemotaxis and granulocyte phagocytosis [89], and the reduction of recruitment, mobilization, growth, and adhesion of circulating angiogenic cells [90, 91]. Moreover, a diminished expression of SDF-1 prevents endothelial progenitor cell recruitment toward the wounds [89]. Therefore, two types of angiogenic conditions coexist in different tissues in diabetes: both anti and pro-angiogenic. This angiogenic paradox indicates that the local microenvironment of each organ definitely plays a very important role in the angiogenic process [83].

3 Hh Generalities

Hh family proteins are morphogens widely distributed throughout much of the animal kingdom being first identified in *Drosophila melanogaster* [92]. There are three mammalian Hh genes named Shh, after a popular video game character; desert hedgehog (Dhh), after an Egyptian species of Hh (Hemiechinus auritus);

and Indian hedgehog (Ihh), a Hh species endemic in Pakistan (Hemiechinus micropus) [14, 15, 93, 94]. Although different features of the molecular processes by which Hh binding to cells stimulates cellular fate have been depicted, it is essential to observe that a variety of aspects of Hh pathway remains unclear.

Classically, Shh is synthesized as precursor which undergoes to different reactions during posttranslational processing, leading to its activation and explosion on cells. *Shh* gene encodes precursor polypeptides of approximately 45 kDa that undergo both amino-terminal signal sequence trimming as well as internal proteolysis. Endoproteolytic cleavage [95] produces a 19 kDa amino-terminal segment with which all known signaling activities are associated and a 26 kDa carboxy-terminal product which functions as a cholesterol transferase. The amino-terminal product of cleavage undergoes two lipophilic modifications. It receives a covalent cholesterol adduct [96] and then a fatty acid adduct, usually palmitate [95]. Consequently, these lipid modifications have great and paradoxical influence on Shh protein properties. On the one hand, cholesterol adduct favors retention of Shh on cell surface [97], and on the other hand, both cholesterol and palmitate adducts are essential for protein activity as well as for establishing range of Shh action [98–100].

Due to its great hydrophobic nature, diverse mechanisms have been suggested for Hh long-range effects, three of which are active diffusion through ECM, indirect transmission through secondary signal cascade, and cytonemes through cellular extensions [101–104]. Moreover, lipid modifications allow the organization of soluble Shh multimeric complex able to activate cascade in distant site [99, 100]. Finally, when Shh is carried by microparticles (MPs) (see below for details), it is able to activates its pathway far from synthesis site [105].

Hh signaling is received in target cells by a receptor complex consisting of two membrane proteins: patched (Ptc), a 12-pass transmembrane receptor with two extracellular hydrophilic loops that mediate Hh binding, and smoothened (Smo), a 7-transembrane-domain protein homologous to G-protein-coupled receptors (GPCR), the signal transducer. Most membrane-bound receptors activate downstream signaling upon ligand binding. In contrast, Ptc is repressed by Hh ligand, freeing Smo for downstream signaling [13].

Canonical Smo downstream signaling leads to the activation and nuclear translocation of zinc finger transcription factors glioma-associated oncogenes 1–3 (Gli1, Gli2, Gli3) and upregulation of target genes including *Ptc* and *Gli1*. On the other hand, several noncanonical Hh signals have been described and seem to depend on alternative, non-transcriptional mechanisms. Among these are (1) proposed Src kinase activation, which mediates Hh effects on spinal commissural axonal projections toward the ventral midline; (2) synchronous Ca^{2+} spikes and inositol triphosphate oscillations at the neuronal primary cilium; and (3) glucose uptake through a Ca^{2+}-AMP-activated protein kinase (AMPK), cilia-dependent pathway. Furthermore, in agreement with its GPCR nature, Smo associates with heterotrimeric G_i proteins and stimulates the small GTPases Rac1 and RhoA, leading to fibroblast migration [106].

Despite the great advances, due to its complexity, knowledge of Hh network is still lacking and needs more work to its elucidation.

4 Hh Signaling in Vascular and Cardiac Development

During development, Hh signal pathway is implicated, not only as a morphogen cascade but also in the regulation of cellular proliferation and differentiation. It has been shown that, in mice, Hh signaling is involved in the vascular morphogenesis of the yolk sac, the first site of blood and blood vessel formation in the mouse. By using embryonic stem cells deficient in Smo, Byrd et al. [107] have shown that in the absence of Hh signaling, embryoid bodies arrest at an earlier stage, and their yolk sacs exhibit severe vascular defects including dense clusters of the vessels packed with hematopoietic cells and failed angiogenesis. Conditional gene targeting in mice helped to demonstrate that cardiomyoblast and perivascular cells are the relevant targets of Hh signaling which regulates the development of distinct vascular subtypes. Whereas Hh signaling to the cardiomyoblast is required for coronary vein development, in the perivascular cells, Hh signaling specifically leads to coronary artery development [108]. Moreover, the role of Hh signaling in vascular development is highly conserved as demonstrated in other models such as avian embryos and zebra fish. Indeed, in avian embryos, the loss or inhibition of Hh signaling results in the abolishment of vascular tube formation, while the implantation of beads containing Shh into endodermless embryos is sufficient to rescue vascular tube formation [109]. In zebra fish embryos carrying null mutations for Shh, endothelial cells failed to undergo arterial differentiation, as defined by the expression of artery-specific markers, such as ephrin-B2a, while injection of mRNA encoding Shh can induce ectopic vascular expression of ephrin-B2a [110]. Interestingly, in this model, VEGF was able to rescue arterial differentiation in the absence of Shh signaling, indicating that during embryonic vascular development VEGF acts downstream of Shh to determine arterial cell fate [110]. In agreement with these findings, other authors have described that Hh is required for VEGF expression during mouse vasculogenesis and that both were required for artery/vein identity in addition to their role in endothelial tube formation [111]. However, hyperactivation of the Hh signaling pathway through deletion of Ptc1 caused increased aorta diameter and increased expression of the arterial Notch ligand Dll4, independently of VEGF levels indicating that Hh signaling modulates distinct vascular patterning events in mouse embryos through both VEGF-dependent and VEGF-independent mechanisms [111].

Very recently, it has been proposed that both Shh and VEGF gradients are important for the randomized distribution of arterial and venous progenitors within the axial vessels in embryonic zebra fish; thus, medial angioblasts located near to the midline of the somites are likely exposed to higher concentration of Shh and VEGF leading to arterial, and not venous, cell differentiation [112].

In addition to participate in the vascular development, it is interesting to note that Hh is involved in the morphogenesis of the heart. Of the point of vu of the embryology, the development of the heart starts very early when progenitor cells in the mesoderm form the primitive cardiac tube. The straight tube then undergoes rightward looping, and subsequently, the atrial and ventricular chambers appear and mature. Due to the physiological function of Hh in determining normal left–right asymmetry, heart tube looping is critically Hh-dependent process (for review see [113]).

Regarding heart development, Hh is, among several factors, an essential coordinator of early cardiomyogenesis [114]. Inhibition of Smo with cyclopamine blocked differentiation of mouse stem cell line P19.CL6 into beating cardiomyocytes [115].

Since Hh is required for normal left–right development, loss of Shh causes several cardiac abnormalities in mice. Embryos from Shh-deficient mice displayed severe reduction of the right ventricle, a prolonged left ventricle, and an extended apex pointing toward the right, septation defects, outflow tract shortening, and dilated inflow tract [114, 116]. Also, aortic arch artery development is abnormal [114]. Interestingly, mice that lack smoothened and mice lacking both Shh and Ihh exhibit more severe cardiac defects such as aberrant cardiac morphogenesis and reduced heart size than mice lacking only Shh [117]. Remarkably, in patients with Smith–Lemli–Opitz syndrome, having a diminished Hh pathway activity due to a deficiency of 7-dehydrocholesterol reductase leading to an improper sterolation of Hh protein [118], congenital heart disorders similar to the defects described in Shh-deficient mice are recurrent.

Peng et al. [119] have recently reported a novel population of multipotent cardiopulmonary progenitors regulated by Hh expression that coordinates heart and lung co-development since Shh−/− mutant mice have a disorganized vascular endothelial plexus that fails to connect to the inflow and outflow tracts of the heart. Also, these authors have shown that inactivation of Smo in cardiopulmonary progenitors inhibited their contribution to airway and vascular smooth muscle at E13.5 in mice. Previous studies have described that noncardiac cells are the source of the Hh signal required for atrial septation [120, 121]; in this way, it has been demonstrated that atrial septum and pulmonary trunk progenitors migrate into the heart after receiving Hh signaling at the second heart field [121].

5 Hh Signaling and New Vessel Formation

In addition to its function in de novo vascularization during embryonic life, Hh pathway plays an active role in postnatal physiology being central in inducing vessel formation. However, the mechanism implicated remains controversial (Fig. 2). Whereas some authors propose that Hh protein can act directly on angiogenic cells (progenitor and mature endothelial cells), other works suggest the indirect effect of Hh to induce angiogenesis rather than a direct effect on

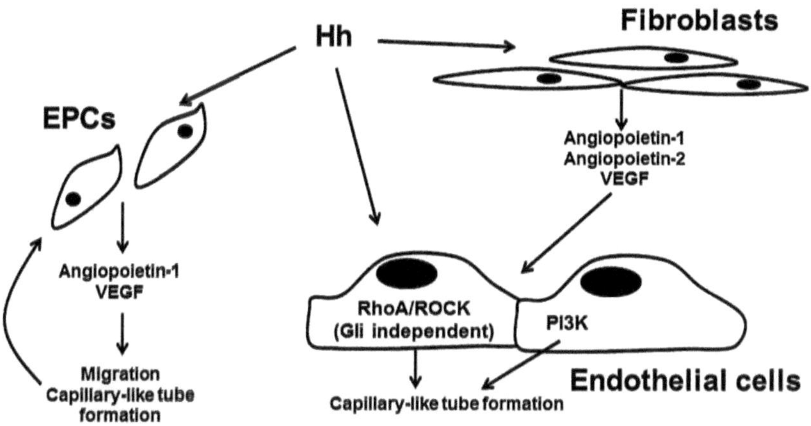

Fig. 2 Direct and indirect Hh signaling in vascular cells. Hedgehog (Hh) protein can act directly on endothelial progenitor cells (EPCs), which secrete pro-angiogenic factors (angiopoietin-1 and VEGF) and favor migration and capillary-like tube formation. On the other hand, Hh stimulates production of angiogenic modulators (angiopoietin-1 and angiopoietin-2 and VEGF) from fibroblasts, which in turn activate phosphatidylinositol 3-kinase (PI3K) pathway on endothelial cells. Moreover, in the same cells, but independently from Gli activation, Hh protein acts on RhoA/ ROCK axis. Both mechanisms lead to the assembling of capillary-like structures

endothelial cells because there was no direct action of Shh on cellular responses of cultured endothelial cells, such as proliferation, migration [2, 122], and serum-deprived survival [2]. Nevertheless, susceptible cells to be activated by Hh must necessarily express the Hh receptor. In this respect, adult cardiac and vascular tissues express Ptc1 and can be stimulated with exogenous administration of Hh as well as by engineered modifications of Hh signaling, that is, by genetic, cellular, or extracellular vesicle tools.

5.1 Implicated Mechanisms in the Direct Effects of Hh on New Vessel Formation: Effects on Stem/Progenitor Cells

Also, overexpression of Hh in mesenchymal stem cells using an Hh plasmid leads to upregulation of Ptc1 that was associated with the increase of expression of secretable angiogenic growth factors including Ang1 and VEGF as well as enhanced migration and capillary-like tube formation [123]. The mechanism implicated the Hh-dependent activation of the inducible nitric oxide synthase (iNOS)/ netrin-1/protein kinase C (PKC) signaling. More interestingly, Hh-transfected mesenchymal stem cells improved blood vessel density in the infarcted heart, suggesting that reprogramming of stem cells with Hh by genetic bioengineering improves their angiogenic potential and could be a promising strategy for the treatment of an infarcted myocardium (see below) [123]. Also, in corneal and ischemic hindlimb models of angiogenesis, Hh induces robust neovascularization. This process is characterized by formation of functional vascular network composed of several venous structures with arteriovenous shunts.

Very recent data showed that Hh pathway was able to induce corneal endothelial cell proliferation. More in detail, leucine-rich repeat G-protein-coupled receptor 5 (LGR5) was the target molecule of Hh signaling in an in vitro model of human corneal endothelial cells displaying stem/progenitor cell characteristics, and their fate was maintained, at least in part, by the activation of Hh pathway [124].

Increased pro-angiogenic ability of endothelial progenitor cells can be acquired by exogenous treatment with recombinant Shh protein or with Shh plasmids. Indeed, recombinant Shh protein (0.01–10 μg/mL) promoted an angiogenic phenotype of endothelial progenitor cells derived from bone marrow of mice, including proliferation and migration and VEGF production. All of these processes are requisite steps in new vessel formation. The mechanisms underlying these effects included activation of PI3K/Akt pathway and were not completely dependent on VEGF [125]. Also, the use of gene therapy in order to increase neovascularization is steadily increasing. Podolska et al. [126] have shown that both intradermal and intramuscular administrations of human Shh plasmids induced increased endothelial progenitor cell recruitment that was evaluated by the enhancement in the skin and muscles of expression of VEGFR-2, CD34, and CD133, three characteristic molecules of endothelial progenitor cell phenotype [126].

5.2 Implicated Mechanisms in the Direct Effects of Hh on New Vessel Formation: Effects on Mature Endothelial Cells

Addition of Shh to cultured endothelial cells causes the formation of vascular network-like structures [109, 122]. The mechanism implicated both PI3K activity and transcriptional regulation pathways in endothelial cells, although PI3K cascade is not involved in Shh-induced Gli 1 nuclear translocation [122].

Recently, in a very elegant study, it has been shown that all three proteins of the Hh family triggered pro-angiogenic responses in human umbilical vein endothelial cells via noncanonical pathways [127]. Thus, Shh activated the small GTPase RhoA in endothelial cells and stimulated formation of capillary-like structures by a Smo- and Gi protein-dependent but Gli- and Ptc-independent pathway. Interestingly, Hh proteins exerted an antiapoptotic effect on endothelial cells by reducing caspase activation and promoting inactivation of the Ptc1 proapoptotic activity, independently of Smo modulation [127]. Altogether these data suggest a dual signaling of Hh proteins; whereas Hh proteins prime the initial steps of angiogenesis, mostly cytoskeletal activation, cell elongation, and extended survival by acting through noncanonical signals, the indirect effects of Hh on Gli-dependent secretion of

VEGF and Ang1 and 2 by mesenchymal cells may provide the additional contribution required to enhance endothelial cell proliferation, migration, survival, and vessel maturation [127]. Similar mechanism implicating RhoA-/Rho-associated protein kinase (ROCK) pathway has been described in a co-culture model of brain microvascular endothelial cells and astrocytes during oxygen–glucose deprivation [128]. In fact, after oxygen–glucose deprivation, astrocytes released Shh which directly promoted angiogenesis via the RhoA/ROCK pathway. However, this was not the sole mechanism, because increased angiogenesis induced by activated astrocytes was not completely reversed in the presence of either cyclopamine or a blocking antibody of Smo.

In vitro treatment with 10 ng/mL Shh-induced human umbilical vein endothelial cell migration and capillary morphogenesis by a noncanonical Shh pathway independent of Gli expression [129]. These authors confirm the previous results of Chinchilla et al. [127] in which Shh-induced angiogenesis was mediated by activation of Rho/ROCK pathway.

Additionally, Shh participates in the stabilization of growing blood vessels by regulating both migration and recruitment of mural cells. This pathway implicated PDGF secreted by VEGF-stimulated endothelial cells which in its turn induced Shh expression in mural cells [130].

5.3 *Involved Mechanisms in the Indirect Effects of Hh on Angiogenesis*

When administered to aged mice undergoing unilateral, surgically induced hindlimb, intramuscular injections of 1 mg/kg Shh–mIgG1 fusion protein was able to induce robust neovascularization of ischemic hindlimb leading to limb salvage. In addition to increase the number of new blood vessels, Hh-induced angiogenesis was characterized by a substantial increase in vessel diameter. In a recent study, it has been showed that after hindlimb ischemia, the process of myogenesis was required for angiogenesis. A key regulator of myogenesis is the downstream Hh transcription factor Gli3, which is essential for the expression of the pro-angiogenic factor thymidine phosphorylase (TYMP) and Ang1, known to promote endothelial cell proliferation and vessel maturation, respectively [131]. In another model of angiogenesis such as the murine corneal angiogenesis model, neovasculature induced by Hh pellets consisted of large, branching vessels that were longer than those generated with VEGF pellet treatment. Histological analysis showed that Hh-activated cells were not endothelial cells (CD31 negative) or periendothelial cells (α-smooth muscle actin negative), but were consistent with interstitial fibroblasts (vimentin positive) surrounding the neovessels. These results suggest that Hh is an indirect angiogenic agent by acting specifically in mesenchymal cells in which Hh induces upregulation of two families of angiogenic growth factors, including VEGF and the Ang1 and Ang2 [132]. The same authors described similar effects in young mice [2]. Moreover, the activation of the Hh signaling is central for the overall production of VEGF, and the related angiogenic response since inhibition of Hh by using a blocking Hh antibody decreased both VEGF production and angiogenesis [2].

Also, in injured skeletal muscle by mechanical crush, it has been shown that a strong Hh-positive signal was detected in skeletal muscle fibers surrounding the injured area, indicating that surviving skeletal muscle fibers are responsible for Hh production following injury [133]. Furthermore, the Hh signaling inhibitor cyclopamine reduced both VEGF and SDF-1 upregulation in injured muscles of mice that was associated with decreased capillary density, impaired skeletal muscle repair, and functional recovery [133].

6 Use of Hh Signaling in Regenerative Medicine

In view of findings illustrated above concerning the key effects of Hh signaling on the new vessel formation, numerous groups have tried to use exogenous Hh in order to repair injured tissues. Two main injuries associated with failed number of functional blood vessels have been studied: myocardial infarction and diabetes.

The ability of Hh to induce tissue repair has been evaluated by the administration of recombinant proteins, gene transfer, or cell therapy. In addition, other studies showed Hh involvement in new vessel formation using extracellular vesicles as vectors.

6.1 Myocardial Infarction

In myocardial infarction, the heart is characterized by the loss of cardiomyocytes, scar formation, ventricular remodeling, and ultimately heart failure. Current pharmacological and interventional approaches are used as bases to prevent loss of cardiac tissue, to enhance angiogenesis, and to reduce left ventricular remodeling (for review see [134]). Thus, Hh signaling activation may represent an excellent candidate to favor angiogenesis.

Although direct injection of recombinant Shh protein restored blood flow in a critical hindlimb ischemia model [132] and reduced myocardial infarction size [135], the limitation of this therapeutic approach is based on short morphogen half-lives in the body [136]. Other groups have also employed either gene therapy [137] or cell therapy [123] in order to enhance Hh signaling with the objective to improve cardiac function following myocardial infarction. However, both of these approaches carry high risk of inducing tumor formation, and thus, trials have not progressed beyond large animal models. Finally, some investigations have focused an approach consisting in a coacervate delivery system of Shh that provides stabilization and protection of the protein and can release it slowly to maintain a constant local concentration within the therapeutic range [138].

Intramyocardial transfer (100 μg/mice) of naked DNA encoding human Shh triggers pleiotropic beneficial effects during myocardial ischemia in adult mice [137]. Thus, Shh gene therapy upregulated Shh pathway in fibroblasts and cardiomyocytes showing its dual direct and indirect effects. Among the most spectacular effects, phShh induced myocardial tissue preservation and repair, not only by promoting neovascularization by increasing endothelial progenitor cell recruitment but also by preserving left ventricular function and by reducing fibrosis and cardiac apoptosis [137].

In order to enhance the pro-angiogenic potential of human circulating progenitor cells (CD34+ cells), a combined model of gene and cell therapy by using Shh-modified human CD34+ cells has been recently described [139]. When injected into the border zone of mice following myocardial infarction, Shh-modified CD34+ cells were protected against ventricular dilation and cardiac functional declines associated with myocardial infarction, reduced infarct size, and increased border zone capillary density. As previously described by our team using T cells [140–144], Shh-modified CD34+ cells are able packaged Shh protein in extracellular vesicles (see below) which is transferred to recipient/target cells and likely participates to beneficial effects induced by Shh-modified CD34+ cells.

6.2 *Diabetes*

Macro- and microvascular complications are the most common major clinical problem of diabetic patients. In particular, patients with diabetes develop microvascular complications such as retinopathy and nephropathy, which can cause blindness and renal failure and also peripheral artery disease leading to limb amputation. It is well accepted that early detection of peripheral artery disease and the subsequent interventions resulting in revascularization will reduce amputations (for review see [145, 146]). Dysfunctional angiogenesis associated with diabetes can be associated with impairment of endothelial function as result of reduced secretion of VEGF and fibroblast growth factor (FGF) and increased oxidative stress as well as failed endothelial progenitor cell function (mobilization, homing, and endothelium repair). Altogether these events reduce tissue vascularization leading to delayed wound healing in diabetic patients [147], and consequently, the improvement of failed functions of endothelial mature cells and endothelial progenitor cells during diabetes represents the goal of future therapeutic approaches.

As during myocardial infarction, several approaches have been used in diabetic animal models in order to attempt to enhance Hh signaling to improve failed angiogenesis. Indeed, local application of 20 μg/mL Hh improved wound healing in streptozotocin-induced diabetic mice by increasing NO production in skin [148]. It has very recently shown that in the same model of streptozotocin-induced diabetes, trombospondin-1 contributed critically to Shh deficiency resulting in dysfunction of bone marrow-derived progenitor cells, and the treatment with recombinant mouse Hh (20 μg/mL) improved the ability of diabetic bone marrow-derived progenitor cells to form capillary-like tubes [149]. This study provides a new potential target in the future angiogenesis-based therapy. Also, subcutaneous injections (1.0 mg/kg, 3 times per week for 4 weeks) of constructed human Shh-rat IgG fusion proteins to increase Shh half-life were able to improve nerve blood flow in streptozotocin-treated rats displaying diabetic neuropathy [150].

Moreover, in diabetic *db/db* mouse model, topical gene therapy with human plasmid Shh (100 μg/mouse) activated Shh pathway accounting for the beneficial effect on microvascular remodeling during cutaneous wound healing. In particular, dermal fibroblasts were activated by exogenous Shh stimulation, resulting in the expression of several angiogenic cytokines and induction of proliferative activity. The effects of Shh on wound healing were associated with enhanced vascularity, with sprouting toward the center part of the wound as well as VEGF upregulation resulting in the formation of mature vessels. In addition to these indirect effects on fibroblasts, Shh increased directly proliferation, adhesion, migration, and tube formation of endothelial progenitor cells [151]. This study suggests that Shh-associated gene therapy mitigates complications of failed angiogenesis in diabetes. Beneficial effects of Shh gene therapy in an experimental model of peripheral ischemia have been more recently described in young mice [17]. Moreover, although Shh gene therapy has been shown to promote ischemia-induced angiogenesis in aged mice [132], a recent study reported that activation of Hh signaling observed in the setting of ischemic muscle repair was impaired in middle-aged and old mice because Gli1 is not upregulated [152, 153]. Also, it has been evidenced that two crucial Hh pathway elements, the ligand Dhh and the G-coupled receptor Smo, were downregulated in aged mice [153]. Therefore, these results indicate that Hh therapy might be helpful to treat aged patient with peripheral arterial diseases. Also, intramuscular treatment with 20–40 μg plasmid encoding the Shh human gene (phShh) induced robust and diffuse activation of the Hh pathway in the skeletal muscle of mice and, interestingly, increased blood flow and both capillary and arteriole densities in mouse ischemic hindlimb. Among cells implicated in angiogenesis, Shh gene therapy increased the number of circulating bone marrow-derived endothelial precursors which contribute to the process of neovascularization. These findings suggest that Shh gene therapy may stimulate several aspects of neovascularization such as mobilization and homing of endothelial progenitor cells during peripheral ischemia and pro-angiogenic factor production. This notion was highlighted by the fact that combined treatment with Shh gene transfer and exogenous bone marrow-derived endothelial progenitor cells was more effective to promote angiogenesis and muscle regeneration than either treatment individually [154].

Also, the Shh agonist SAG1.3 at 5 mg/kg/day for 21 days significantly restored Ptc1 and Gli1 protein levels in streptozotocin-treated mice with myocardial infarction. In addition, same dose of SAG1.3 enhanced capillary density, reduced the percentage myocardial infarct, and then improved cardiac function in these mice [135] suggesting that therapeutic strategies based on Shh agonists may provide an effective means for ameliorating diabetic cardiac dysfunction.

7 Extracellular Vesicles and Shh Signaling

7.1 Extracellular Vesicle Generalities

Extracellular vesicles such as exosomes and MPs are small vesicles (0.01–1 μm) surrounded by the phospholipid bilayer and released by cells. While exosomes originate from ectocytosis of multivesicular bodies, MPs are generated by direct budding from the plasma membrane (Fig. 3). Extracellular vesicles express antigenic profile characteristic of the cell they originate. Although for years the extracellular vesicles have been considered inert dust without specific function, not long ago they are deemed as vectors by actively orchestrating important physiological functions and pathophysiological processes. The mechanisms by

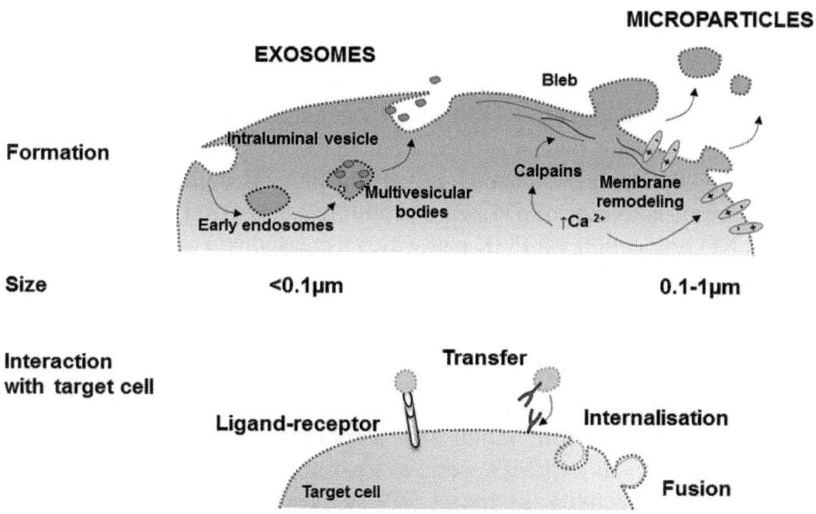

Fig. 3 Extracellular vesicles (EVs): exosomes and microparticles (MPs). EVs are small membrane-limited vesicles, generated from cells and released in the extracellular space. They include both exosomes and MPs which differ in their origin and size. Exosomes are vesicles smaller than 0.1 μm which originated from internal vesicles of an endosomal compartment, the multivesicular body, and liberated after exocytic fusion of this organelle with the plasma membrane. However, MPs are vesicles generated from membrane remodeling and blebbing occurring after cell activation or apoptosis. Their size is comprised between 0.1 and 1 μm. Although exosomes and MPs are different in composition, they deliver to the target cell biological information in the same manners by ligand–receptor interaction, by transfer of bioactive molecules, by internalization, or by fusion

which extracellular vesicles mediate intercellular communication are various [155]. Among them, extracellular vesicles can bear single or combinations of ligands that would simultaneously engage different cell-surface receptors. All these evidences sustain the idea of extracellular vesicles as vectors able to carry transcellular messages and thus allow considering them as authentic players in the regulation of physiological and pathological processes. Particular attention had been paid to this novel method of cell communication, because extracellular vesicles are able to exchange biological message between different kinds of cells, not only in the proximity area of their generation but also far from cells they stem from. Hemostasis and thrombosis, diabetes, inflammation, atherosclerosis, angiogenesis, tumor progression, apoptosis, vascular cell proliferation, and outgrowth of transplanted hematopoietic stem cells [156–158] represent some physiological and pathological situations where generation of extracellular vesicles is enhanced and subsequently may be involved in altered angiogenesis. In fact, it has been shown that different stages of vessel formation are regulated by extracellular vesicles.

Depending on stimulation, cells generate MPs with specific functional effects; in fact, it has been shown that MPs shed by apoptotic lymphocytes (LMPs), inhibited in vitro and in vivo angiogenesis, by suppressing vascular cell survival, proliferation, and migration. They induced ROS production which occurs upstream of induction of CD36 (thrombospondin [TSP]1 receptor) with subsequent suppression of VEGF/VEGFR2 signaling pathway. Also, LMPs were able to antagonize in vivo neovascularization induced by VEGF [159]. Indeed, the antiangiogenic effects of LMPs lead to a significant reduction of pathological retinal angiogenesis through modulation of VEGF signaling [160]. LMPs markedly inhibited choroidal angiogenesis via mechanisms that were dependent on the integrity of the retinal pigment epithelium and that were mediated largely by the pigment epithelium-derived factor and proapoptotic activities of p75 neurotrophin receptor [161]. Furthermore, LMPs decreased NO production via PI3K pathway. Decreased NO generation was associated with enhanced phosphorylation of eNOS on its inhibitory site and overexpression of caveolin-1 [162]. Indeed, antiangiogenic effects of LMPs were linked to oxidative stress and reduced NO release from endothelial cells [163]. Furthermore, it has been shown that these LMPs inhibited lung tumor growth and microvessel density and limited local production of VEGF. Decreased lung carcinoma cell viability and proliferation were associated with an increased apoptosis. These antitumor effects of LMPs were, at least in part, dependent on low-density lipoprotein receptor activity. In vivo, LMPs suppressed tumor growth by targeting tumor angiogenesis and cell growth by interfering with the VEGF pathway. This finding allows to hypothesize that LMPs transfer into tumor cells may suppress pro-angiogenic and pro-growth response under pathophysiological conditions [164].

All these studies showed that MPs act along all process of vessel formation by modulating cell function, differentiation, proliferation, migration, and adhesion. Therefore, they also may serve as potential target or tools in therapeutic approaches.

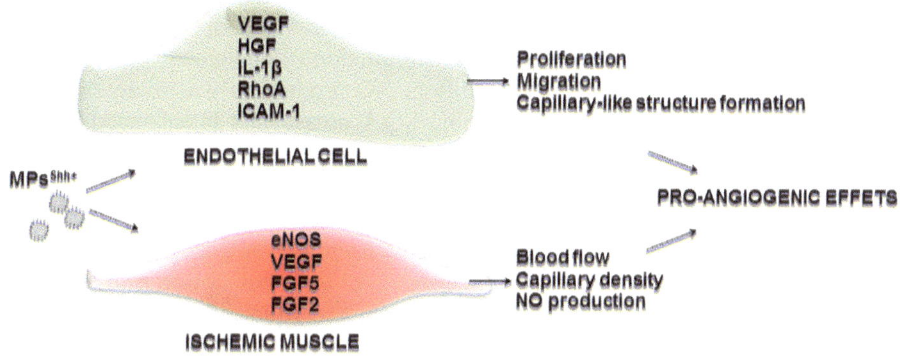

Fig. 4 Pro-angiogenic effects of microparticles (MPs) carrying sonic hedgehog (MPs^{Shh+}). MPs^{Shh+} exhibit pro-angiogenic effects in both in vitro and in vivo models. They modulate endothelial cell proliferation, migration, and adhesion and induce formation of capillary-like structures, by acting on regulation of vascular endothelial growth factor (VEGF), hepatocyte growth factor (HGF), interleukin 1 beta (IL-1β), RhoA, and intracellular adhesion molecule-1 (ICAM-1) expression. Moreover, MPs^{Shh+} favor postischemic neovascularization by enhanced expression of different angiogenic modulators, including the endothelial isoform of nitric oxide synthase (eNOS), VEGF, the isoforms of fibroblast growth factor (FGF5 and 2) leading to improved blood flow, capillary density, as well as nitric oxide (NO) production

7.2 *Use of Extracellular Vesicles as Cargo for Shh Signaling*

We have generated engineered MPs through the activation of T lymphocytes with phytohemagglutinin and phorbol ester and further induction of apoptosis with actinomycin D. The generated MPs expressed Shh (MPs^{Shh+}) [105] and displayed pro-angiogenic effects on target cells [141]. Indeed, in vitro treatment of human umbilical vein endothelial cells with MPs^{Shh+} induced formation of capillary-like tubes of these cells, mainly the increase of cell proliferation and endothelial adhesion through the increase of intercellular adhesion molecule-1 expression. These effects involved Rho kinase pathway and paradoxically are inhibited both by blocking Smo activity with cyclopamine and activating Hh pathway by removing Ptc antagonist effect. These results suggest that noncanonical mechanisms are implicated in the MPs^{Shh+} effects and might involve other putative Hh receptors. In addition, MPs^{Shh+} enhanced mRNA and protein expressions of pro-angiogenic factors such as hepatocyte growth factor, VEGF, and interleukin-1β indicating that Shh carried by MPs acts on a large number of target genes that regulate angiogenesis (Fig. 4). All these effects were obtained with low concentration of Shh (largely lower than 10 μg/mL, since this was the total concentration of MP proteins used).

In addition, under pathophysiological conditions, MPs^{Shh+} were able to favor postischemic neovascularization. Especially in vivo treatment of hindlimb ischemic mice with MPs^{Shh+} improved blood flow in the ischemic leg as well as the number of capillaries when compared with the nonischemic leg [142, 143]. Interestingly, the improvement of the blood flow was the result of the interaction between ligand Shh carried by MPs and its receptors and was evidenced by blocking of Shh receptor with cyclopamine, as well as the stimulation of endogenous Shh expression on mouse ischemic muscle. In addition, NO production was increased, as a consequence of the activation of endothelial NO-synthase, and an enhanced expression of several pro-angiogenic factors, including FGF5, FGF2, and VEGF, was reported in the muscle from the ischemic leg after MPs^{Shh+} treatment (Fig. 4). Altogether, these results suggest that MPs^{Shh+} are potent regulators of angiogenesis during muscle regeneration after ischemia in mice and they might be potential tools in diseases associated with failed angiogenesis. As described above, other authors have also shown that extracellular vesicles generated from Shh-modified CD34+ cells are able to transfer Shh signaling to recipient/target cells and likely participate to beneficial effects induced by Shh-modified CD34+ cells [139].

However, in other models such as chronic liver diseases, angiogenesis is a pathological hallmark and the signal carried by MPs^{Shh+} could be deleterious. Indeed, it has been shown that in cirrhotic livers, myofibroblastic hepatic stellate cells and cholangiocytes released extracellular vesicles containing biologically active Hh ligands [165]. These vesicles induced changes in gene expression of hepatic sinusoidal endothelial cells such as an increase of iNOS and CD31 expressions associated with Gli2 expression confirming that the changes that occur in these cells after treatment with vesicles carrying Shh result from the activation of Hh pathway. These data suggest a novel mechanism for cirrhotic vasculopathy implicating vesicles carrying Shh released from Hh-producing liver cells that accumulate in fibrotic livers and promote remodeling of the hepatic sinusoids during cirrhosis.

8 Conclusions and Future Directions

In view of these findings, activation of Hh signaling may represent an interesting approach to palliate vascular alterations associated with failed angiogenesis. However, in order to reduce possible side effects of Hh treatment, a local application may be used rather than a systemic application. Other problem to overcome would be the replacement of Hh-associated plasmid administration by other safer systems. Finally, the development of new Hh agonists with more specific activity on failed vascular (non-tumor) angiogenesis than on tumor angiogenesis should be an interesting challenge.

References

1. Fischer C, Schneider M, Carmeliet P. Principles and therapeutic implications of angiogenesis, vasculogenesis and arteriogenesis. Handb Exp Pharmacol. 2006;(176 Pt 2):157–212

2. Pola R, Ling LE, Aprahamian TR, Barban E, Bosch-Marce M, Curry C, Corbley M, Kearney M, Isner JM, Losordo DW (2003) Postnatal recapitulation of embryonic hedgehog pathway in response to skeletal muscle ischemia. Circulation 108:479–485
3. Machold R, Hayashi S, Rutlin M, Muzumdar MD, Nery S, Corbin JG, Gritli-Linde A, Dellovade T, Porter JA, Rubin LL (2003) Sonic hedgehog is required for progenitor cell maintenance in telencephalic stem cell niches. Neuron 39:937–950
4. Adolphe C, Narang M, Ellis T, Wicking C, Kaur P, Wainwright B (2004) An in vivo comparative study of sonic, desert and Indian hedgehog reveals that hedgehog pathway activity regulates epidermal stem cell homeostasis. Development 131:5009–5019
5. Watkins DN, Berman DM, Burkholder SG, Wang B, Beachy PA, Baylin SB (2003) Hedgehog signalling within airway epithelial progenitors and in small-cell lung cancer. Nature 422:313–317
6. Bijlsma MF, Spek CA, Peppelenbosch MP (2004) Hedgehog: an unusual signal transducer. Bioessays 26:387–394
7. Lum L, Beachy PA (2004) The Hedgehog response network: sensors, switches, and routers. Science 304:1755–1759
8. Hooper JF, Scott MP (2005) Communicating with Hedgehogs. Nat Rev Mol Cell Biol 6:306–317
9. Ruiz i Altaba A, Mas C, Stecca B (2007) The Gli code: an information nexus regulating cell fate, stemness and cancer. Trends Cell Biol 17:438–447
10. Dellovade T, Romer JT, Curran T, Rubin LL (2006) The Hedgehog pathway and neurological disorders. Annu Rev Neurosci 29:539–563
11. Taipale J, Beachy P (2001) The Hedgehog and Wnt signalling pathways in cancer. Nature 411:349–354
12. Pasca di Magliano M, Hebrok M (2003) Hedgehog signalling in cancer formation and maintenance. Nat Rev Cancer 3:903–911
13. Briscoe J, Therond P (2013) The mechanisms of Hedgehog signalling and its roles in development and disease. Nat Rev Mol Cell Biol 14:416–429
14. Katoh Y, Katoh M (2005) Comparative genomics on Sonic hedgehog orthologs. Oncol Rep 14:1087–1090
15. Katoh Y, Katoh M (2005) Hedgehog signaling in gastric cancer. Cancer Biol Ther 4:1050–1054
16. Kasper M, Regl G, Frischaf AM, Aberger F (2006) Gli transcription factors: mediators of oncogenic Hedgehog signalling. Eur J Cancer 42:437–445
17. Piccioni A, Gaetani E, Neri V, Gatto I, Palladino M, Silver M, Smith RC, Giarretta I, Pola E, Hlatky L, Pola R (2014) Sonic hedgehog therapy in a mouse model of age-associated impairment of skeletal muscle regeneration. J Gerontol A Biol Sci Med Sci 69:245–252
18. Griffioen AW, Molema G (2000) Angiogenesis: Potentials for pharmacologic intervention in the treatment of cancer, cardiovascular diseases, and chronic inflammation. Pharmacol Rev 52:237–268
19. Noden DM (1989) Embryonic origins and assembly of blood vessels. Am Rev Respir Dis 140:1097–1103
20. Murohara T (2001) Therapeutic vasculogenesis using human cord blood-derived endothelial progenitors. Trends Cardiovasc Med 11:303–307
21. Ema M, Rossant J (2003) Cell fate decisions in early blood vessel formation. Trends Cardiovasc Med 13:254–259
22. Coultas L, Chawengsaksophak K, Rossant J (2005) Endothelial cells and VEGF in vascular development. Nature 438:937–945
23. Jain RK (2003) Molecular regulation of vessel maturation. Nat Med 9:685–693
24. Ausprunk DH, Folkman J (1977) Migration and proliferation of endothelial cells in preformed and newly formed blood vessels during angiogenesis. Microvasc Res 14:53–65
25. Risau W (1997) Mechanisms of angiogenesis. Nature 386:671–674

26. Drake CJ, Cheresh DA, Little CD (1995) An antagonist of integrin avβ3 prevents maturation of blood vessels during embryonic neovascularization. J Cell Sci 108:2655–2661
27. Auerbach HR, Auerbach W (1997) Profound effects on vascular development caused by perturbations of during organogenesis. Am J Pathol 151:1183–1186
28. Pardanaud L, Dieterlen-Lièvre F (1999) Manipulation of the angiopoietic/hemangiopoietic commitment in the avian embryo. Development 26:617–627
29. Asahara T, Murohara T, Sullivan A, Silver M, Zee RVD, Li T, Witzenbichler B, Schattemen G, Isner JM (1997) Isolation of putative progenitor endothelial cells for angiogenesis. Science 275:964–967
30. Milkiewicz M, Ispanovic E, Doyle JL, Haas TL (2006) Regulators of angiogenesis and strategies for their therapeutic manipulation. Int J Biochem Cell Biol 38:333–357
31. Pepper MS (2001) Role of the matrix metalloproteinase and plasminogen activator-plasmin systems in angiogenesis. Arterioscler Thromb Vasc Biol 21:1104–1117
32. Axnick J, Lammert E (2012) Vascular lumen formation. Curr Opin Hematol 19:192–198
33. Djonov V, Schmid M, Tschanz SA, Burri PH (2000) Intussusceptive angiogenesis: its role in embryonic vascular network formation. Circ Res 86:286–292
34. Djonov VG, Kurz H, Burri PH (2002) Optimality in the developing vascular system: branching remodeling by means of intussusception as an efficient adaptation mechanism. Dev Dyn 224:391–402
35. Galis ZS, Khatri JJ (2002) Matrix metalloproteinases in vascular remodeling and atherogenesis: the good, the bad, and the ugly. Circ Res 90:251–262
36. Glagov S, Weisenberg E, Zarins CK, Stankunavicius R, Kolettis GJ (1987) Compensatory enlargement of human atherosclerotic coronary arteries. N Engl J Med 316:1371–1375
37. Rossant J, Howard L (2002) Signaling pathways in vascular development. Annu Rev Cell Dev Biol 18:541–573
38. Maisonpierre PC, Suri C, Jones PF, Bartunkova S, Wiegand SJ, Radziejewski C, Compton D, McClain J, Aldrich TH, Papadopoulos N, Daly TJ, Davis S, Sato TN, Yancopoulos GD (1997) Angiopoietin-2, a natural antagonist for Tie2 that disrupts in vivo angiogenesis. Science 277:55–60
39. Yue PYK, Mak NK, Cheng YK, Leung KW, Ng TB, Fan DTB, Yeung HW, Wong RNS (2007) Pharmacogenomics and the Yin/Yang actions of ginseng: anti-tumor, angiomodulating and steroid-like activities of ginsenosides. Chin Med 2:6
40. Carmeliet P, Ng YS, Nuyens D, Theilmeier G, Brusselmans K, Cornelissen I, Ehler E, Kakkar VV, Stalmans I, Mattot V, Perriard JC, Dewerchin M, Flameng W, Nagy A, Lupu F, Moons L, Collen D, D'Amore PA, Shima DT (1999) Impaired myocardial angiogenesis and ischemic cardiomyopathy in mice lacking the vascular endothelial growth factor isoforms VEGF164 and VEGF188. Nat Med 5:495–502
41. Ferrara N, Alitalo K (1999) Clinical applications of angiogenic growth factors and their inhibitors. Nat Med 5:1359–1364
42. Fadini GP, Losordo D, Dimmeler S (2012) Critical reevaluation of endothelial progenitor cell phenotypes for therapeutic and diagnostic use. Circ Res 110:624–637
43. Ng YS, D'Amore PA (2001) Therapeutic angiogenesis for cardiovascular disease. Curr Contr Trials Cardiovasc Med 2:278–285
44. Tunyogi-Csapo M, Koreny T, Vermes C, Galante JO, Jacobs JJ, Glant TT (2007) Role of fibroblasts and fibroblast-derived growth factors in periprosthetic angiogenesis. J Orthop Res 25:1378–1388
45. Rey S, Semenza GL (2010) Hypoxia-inducible factor-1-dependent mechanisms of vascularization and vascular remodelling. Cardiovasc Res 86:236–242
46. Lau YT, Ma WC (1996) Nitric oxide inhibits migration of cultured endothelial cells. Biochem Biophys Res Commun 221:670–674
47. Tuomisto TT, Rissanen TT, Korkeela A, Korkeela A, Rutanen J, Ylä-Herttuala S (2004) HIF-VEGF-VEGFR-2, TNF-a and IGF pathways are upregulated in critical human skeletal muscle ischemia as studied with DNA array. Atherosclerosis 174:111–120

48. Zachary I (2003) VEGF signalling: integration and multi-tasking in endothelial cell biology. Biochem Soc Trans 31:1171–1177
49. Murohara T, Horowitz JR, Silver M, Tsurumi Y, Chen D, Sullivan A, Isner JM (1998) Vascular endothelial growth factor/vascular permeability factor enhances vascular permeability via nitric oxide and prostacyclin. Circulation 97:99–107
50. Neagoe PE, Lemieux C, Sirois MG (2005) Vascular endothelial growth factor (VEGF)-A165-induced prostacyclin synthesis requires the activation of VEGF receptor-1 and −2 heterodimer. J Biol Chem 280:9904–9912
51. Silvestre JS, Smadja DM, Lévy BI (2013) Postischemic revascularization: from cellular and molecular mechanisms to clinical applications. Physiol Rev 93:1743–1802
52. Wang GL, Jiang BH, Rue EA, Semenza GL (1995) Hypoxia-inducible factor 1 is a basic-helix-loop-helix-PAS heterodimer regulated by cellular O_2 tension. Proc Natl Acad Sci U S A 92:5510–5514
53. Ziello JE, Jovin IS, Huang Y (2007) Hypoxia-Inducible Factor (HIF)-1 regulatory pathway and its potential for therapeutic intervention in malignancy and ischemia. Yale J Biol Med 80:51–60
54. Fukumura D, Kashiwagi S, Jain RK (2006) The role of nitric oxide in tumour progression. Nat Rev Cancer 6:521–534
55. Ohtani K, Dimmeler S (2011) Control of cardiovascular differentiation by microRNAs. Basic Res Cardiol 106:5–11
56. Baiguera S, Ribatti D (2013) Endothelialization approaches for viable engineered tissues. Angiogenesis 16:1–14
57. Hunting CB, Noort WA (2005) Zwaginga JJ Circulating endothelial (progenitor) cells reflect the state of the endothelium: vascular injury, repair and neovascularization. Vox Sang 88:1–9
58. Dimmeler S, Fleming I, Fisslthaler B, Hermann C, Busse R, Zeiher AM (1999) Activation of nitric oxide synthase in endothelial cells by Akt-dependent phosphorylation. Nature 399:601–605
59. Dimmeler S, Hermann C, Galle J, Zeiher AM (1999) Upregulation of superoxide dismutase and nitric oxide synthase mediates the apoptosis-suppressive effects of shear stress on endothelial cells. Arterioscler Thromb Vasc Biol 19:656–664
60. Ballieux BE, Hiemstra PS, Klar-Mohamad N, Hagen EC, van Es LA, van der Woude FJ, Daha MR (1994) Detachment and cytolysis of human endothelial cells by proteinase 3. Eur J Immunol 24:3211–3215
61. Re F, Zanetti A, Sironi M, Polentarutti N, Lanfrancone L, Dejana E, Colotta F (1994) Inhibition of anchorage-dependent cell spreading triggers apoptosis in cultured human endothelial cells. J Cell Biol 127:537–546
62. Mancuso P, Calleri A, Cassi C, Gobbi A, Capillo M, Pruneri G, Martinelli G, Bertolini F (2003) Circulating endothelial cells as a novel marker of angiogenesis. Adv Exp Med Biol 522:83–97
63. Gill M, Dias S, Hattori K, Rivera ML, Hicklin D, Witte L, Girardi L, Yurt R, Himel H, Rafii S (2001) Vascular trauma induces rapid but transient mobilization of VEGFR2(+) AC133(+) endothelial precursor cells. Circ Res 88:167–174
64. Bombeli T, Muller M, Straub PW, Haeberli A (1996) Cyclosporine induced detachment of vascular endothelial cells initiates the intrinsic coagulation system in plasma and whole blood. J Lab Clin Med 127:621–634
65. Ruegg C, Yilmaz A, Bieler G, Bamat J, Chaubert P, Lejeune FJ (1998) Evidence for the involvement of endothelial cell integrin alphaVbeta3 in the disruption of the tumor vasculature induced by TNF and IFN-gamma. Nat Med 4:408–414
66. Asahara T, Kawamoto A (2004) Endothelial progenitor cells for postnatal vasculogenesis. Am J Physiol Cell Physiol 287:C572–C579
67. Werner N, Nickenig G (2006) Influence of cardiovascular risk factors on endothelial progenitor cells: limitations for therapy? Arterioscler Thromb Vasc Biol 26:257–266

68. Peichev M, Naiyer AJ, Pereira D, Zhu Z, Lane WJ, Williams M, Oz MC, Hicklin DJ, Witte L, Moore MA, Rafii S (2000) Expression of VEGFR-2 and AC133 by circulating human CD34 (+) cells identifies a population of functional endothelial precursors. Blood 95:952–958
69. Asahara T, Kawamoto A, Masuda H (2011) Concise review: circulating endothelial progenitor cells for vascular medicine. Stem Cells 29:1650–1655
70. Xu Q (2007) Progenitor cells in vascular repair. Curr Opin Lipidol 18:534–539
71. Lapergue B, Mohammad A, Shuaib A (2007) Endothelial progenitor cells and cerebrovascular diseases. Prog Neurobiol 83:349–362
72. Povsic T, Goldschmidt-Clermont P (2008) Endothelial progenitor cells: Markers of vascular reparative capacity. Ther Adv Cardiovasc Dis 2:199–213
73. Goligorsky MS (2013) Salven P Concise review: endothelial stem and progenitor cells and their habitats. Stem Cells Transl Med 2:499–504
74. Hillebrands JL, Onuta G, Rozing J (2005) Role of progenitor cells in transplant arteriosclerosis. Trends Cardiovasc Med 15:1–8
75. Hillebrands JL, Klatter FA, Bruggeman CA, Rozing J (2001) Development of transplant arteriosclerosis after allogeneic aorta transplantation in the rat: influence of recipient genotype. Transplant Proc 33:324–325
76. Simper D, Wang S, Deb A, Holmes D, McGregor C, Frantz R, Kushwaha SS, Caplice NM (2003) Endothelial progenitor cells are decreased in blood of cardiac allograft patients with vasculopathy and endothelial cells of noncardiac origin are enriched in transplant atherosclerosis. Circulation 15(108):143–149
77. Minami E, Laflamme MA, Saffitz JE, Murry CE (2005) Extracardiac progenitor cells repopulate most major cell types in the transplanted human heart. Circulation 112:2951–2958
78. Takahashi T, Kalka C, Masuda H, Chen D, Silver M, Kearney M, Magner M, Isner JM, Asahara T (1999) Ischemia- and cytokine-induced mobilization of bone marrow-derived endothelial progenitor cells for neovascularization. Nat Med 5:434–438
79. Shintani S, Murohara T, Ikeda H, Ueno T, Sasaki K, Duan J, Imaizumi T (2001) Augmentation of postnatal neovascularization with autologous bone marrow transplantation. Circulation 103:897–903
80. Laufs U, Werner N, Link A, Endres M, Wassmann S, Jürgens K, Miche E, Böhm M, Nickenig G (2004) Physical training increases endothelial progenitor cells, inhibits neointima formation, and enhances angiogenesis. Circulation 109:220–226
81. Jevon M, Dorling A, Hornick PI (2008) Progenitor cells and vascular disease. Cell Prolif 41:146–164
82. Urbich C (2004) Dimmeler S Endothelial progenitor cells: characterization and role in vascular biology. Circ Res 95:343–353
83. Waltenberger J (2007) New horizons in diabetes therapy: the angiogenesis paradox in diabetes: description of the problem and presentation of a unifying hypothesis Immunol. Endocrinol Metab Agents Med Chem 7:87–93
84. Khazaei M, Fallahzadeh AR, Sharifi MR, Afsharmoghaddam N, Javanmard SH, Salehi E (2011) Effects of diabetes on myocardial capillary density and serum angiogenesis biomarkers in male rats. Clinics (Sao Paulo) 66:1419–1424
85. Frank RN (2004) Diabetic retinopathy. N Engl J Med 350:48–58
86. Hammes HP (2011) Diabetic retinopathy and maculopathy. Internist (Berl) 52:518–532
87. Antonetti DA, Klein R, Gardner TW (2012) Diabetic retinopathy. N Engl J Med 366 (13):1227–1239
88. Kumar B, Gupta SK, Saxena R, Srivastava S (2012) Current trends in the pharmacotherapy of diabetic retinopathy. J Postgrad Med 58:132–139
89. Kolluru GK, Bir SC, Kevil CG (2012) Endothelial dysfunction and diabetes: effects on angiogenesis, vascular remodeling, and wound healing. Int J Vasc Med 2012:918267
90. Capla JM, Grogan RH, Callaghan MJ, Galiano RD, Tepper OM, Ceradini DJ, Gurtner GC (2007) Diabetes impairs endothelial progenitor cell-mediated blood vessel formation in response to hypoxia. Plast Reconstr Surg 119:59–70

91. Bento CF, Pereira P (2011) Regulation of hypoxia-inducible factor 1 and the loss of the cellular response to hypoxia in diabetes. Diabetologia 54:1946–1956
92. Nüsslein-Volhard C, Wieschaus E (1980) Mutations affecting segment number and polarity in Drosophila. Nature 287:795–801
93. Marigo V, Roberts DJ, Lee SM, Tsukurov O, Levi T, Gastier JM, Epstein DJ, Gilbert DJ, Copeland NG, Seidman CE, Jenkins NA, Seidman JG, Andrew P, Mcmahon AP, Tabin C (1995) Cloning, expression, and chromosomal location of SHH and IHH: two human homologues of the Drosophila segment polarity gene hedgehog. Genomics 28:44–51
94. van den Brink GR, Bleuming SA, Hardwick JC, Schepman BL, Offerhaus GJ, Keller JJ, Nielsen C, Gaffield W, van Deventer SJ, Roberts DJ (2004) Peppelenbosch MP Indian Hedgehog is an antagonist of Wnt signaling in colonic epithelial cell differentiation. Nat Genet 36:277–282
95. Mann RK, Beachy PA (2000) Cholesterol modification of proteins. Biochim Biophys Acta 1529:188–202
96. Porter JA, Young KE, Beachy PA (1996) Cholesterol modification of hedgehog signaling proteins in animal development. Science 274:255–259
97. Roelink H, Porter JA, Chiang C, Tanabe Y, Chang DT, Beachy PA, Jessell TM (1995) Floor plate and motor neuron induction by different concentrations of the amino-terminal cleavage product of sonic hedgehog autoproteolysis. Cell 81:445–455
98. Lewis PM, Dunn MP, McMahon JA, Logan M, Martin JF, St-Jacques B, McMahon AP (2001) Cholesterol modification of sonic hedgehog is required for long-range signaling activity and effective modulation of signaling by Ptc1. Cell 105:599–612
99. Chen MH, Li YJ, Kawakami T, Xu SM, Chuang PT (2004) Palmitoylation is required for the production of a soluble multimeric Hedgehog protein complex and long-range signaling in vertebrates. Genes Dev 15(18):641–659
100. Chen MH, Li YJ, Kawakami T, Xu SM, Chuang PT (2004) Palmitoylation is required for the production of a soluble multimeric Hedgehog protein complex and long-range signaling in vertebrates. Genes Dev 18:641–659
101. Lee JJ, von Kessler DP, Parks S, Beachy PA (1992) Secretion and localized transcription suggest a role in positional signaling for products of the segmentation gene hedgehog. Cell 71:33–50
102. Basler K, Struhl G (1994) Compartment boundaries and the control of Drosophila limb pattern by hedgehog protein. Nature 368:208–214
103. Struhl G, Barbash DA, Lawrence PA (1997) Hedgehog organises the pattern and polarity of epidermal cells in the Drosophila abdomen. Development 124:2143–2154
104. Zeng X, Goetz JA, Suber LM, Scott WJ Jr, Schreiner CM, Robbins DJ (2001) A freely diffusible form of Sonic hedgehog mediates long-range signalling. Nature 411:716–720
105. Martínez MC, Larbret F, Zobairi F, Coulombe J, Debili N, Vainchenker W, Ruat M, Freyssinet JM (2006) Transfer of differentiation signal by membrane microvesicles harboring hedgehog morphogens. Blood 108:3012–3020
106. Ruat M, Hoch L, Faure H, Rognan D (2014) Targeting of Smoothened for therapeutic gain. Trends Pharmacol Sci 35:237–246
107. Byrd N, Becker S, Maye P, Narasimhaiah R, St-Jacques B, Zhang X, McMahon J, McMahon A, Grabel L (2002) Hedgehog is required for murine yolk sac angiogenesis. Development 129:361–372
108. Lavine KJ, Long F, Choi K, Smith C, Ornitz DM (2008) Hedgehog signaling to distinct cell types differentially regulates coronary artery and vein development. Development 135:3161–3171
109. Vokes SA, Yatskievych TA, Heimark RL, McMahon J, McMahon AP, Antin PB, Krieg PA (2004) Hedgehog signaling is essential for endothelial tube formation during vasculogenesis. Development 131:4371–4380

110. Lawson ND, Vogel AM, Weinstein BM (2002) Sonic hedgehog and vascular endothelial growth factor act upstream of the Notch pathway during arterial endothelial differentiation. Dev Cell 3:127–136
111. Coultas L, Nieuwenhuis E, Anderson GA, Cabezas J, Nagy A, Henkelman RM, Hui CC, Rossant J (2010) Hedgehog regulates distinct vascular patterning events through VEGF-dependent and -independent mechanisms. Blood 116:653–660
112. Kohli V, Schumacher JA, Desai SP, Rehn K, Sumanas S (2013) Arterial and venous progenitors of the major axial vessels originate at distinct locations. Dev Cell 25:196–206
113. Bijlsma MF, Peppelenbosch MP, Spek CA (2006) Hedgehog morphogen in cardiovascular disease. Circulation 114:1985–1991
114. Washington Smoak I, Byrd NA, Abu-Issa R, Goddeeris MM, Anderson R, Morris J, Yamamura K, Klingensmith J, Meyers EN (2005) Sonic hedgehog is required for cardiac outflow tract and neural crest cell development. Dev Biol 283:357–372
115. Clement CA, Kristensen SG, Møllgård K, Pazour GJ, Yoder BK, Larsen LA, Christensen ST (2009) The primary cilium coordinates early cardiogenesis and hedgehog signaling in cardiomyocyte differentiation. J Cell Sci 122:3070–3082
116. Tsukui T, Capdevila J, Tamura K, Ruiz-Lozano P, Rodriguez-Esteban C, Yonei-Tamura S, Magallón J, Chandraratna RA, Chien K, Blumberg B, Evans RM, Belmonte JC (1999) Multiple left-right asymmetry defects in Shh(−/−) mutant mice unveil a convergence of the shh and retinoic acid pathways in the control of Lefty-1. Proc Natl Acad Sci U S A 96:11376–11381
117. Zhang XM, Ramalho-Santos M, McMahon AP (2001) Smoothened mutants reveal redundant roles for Shh and Ihh signaling including regulation of L/R symmetry by the mouse node. Cell 106:781–792
118. DeBarber AE, Eroglu Y, Merkens LS, Pappu AS, Steiner RD (2011) Smith–Lemli–Opitz syndrome. Expert Rev Mol Med 13:e24
119. Peng T, Tian Y, Boogerd CJ, Lu MM, Kadzik RS, Stewart KM, Evans SM, Morrisey EE (2013) Coordination of heart and lung co-development by a multipotent cardiopulmonary progenitor. Nature 500:589–592
120. Goddeeris MM, Rho S, Petiet A, Davenport CL, Johnson GA, Meyers EN, Klingensmith J (2008) Intracardiac septation requires hedgehog-dependent cellular contributions from outside the heart. Development 135:1887–1895
121. Hoffmann AD, Peterson MA, Friedland-Little JM, Anderson SA, Moskowitz IP (2009) Sonic hedgehog is required in pulmonary endoderm for atrial septation. Development 136:1761–1770
122. Kanda S, Mochizuki Y, Suematsu T, Miyata Y, Nomata K, Kanetake H (2003) Sonic hedgehog induces capillary morphogenesis by endothelial cells through phosphoinositide 3-kinase. J Biol Chem 78:8244–8249
123. Ahmed RP, Haider KH, Shujia J, Afzal MR, Ashraf M (2010) Sonic Hedgehog gene delivery to the rodent heart promotes angiogenesis via iNOS/netrin-1/PKC pathway. PLoS One 5: e8576
124. Hirata-Tominaga K, Nakamura T, Okumura N, Kawasaki S, Kay EP, Barrandon Y, Koizumi N, Kinoshita S (2013) Corneal endothelial cell fate is maintained by LGR5 through the regulation of hedgehog and Wnt pathway. Stem Cells 31:1396–1407
125. Fu JR, Liu WL, Zhou JF, Sun HY, Xu HZ, Luo L, Zhang H, Zhou YF (2006) Sonic hedgehog protein promotes bone marrow-derived endothelial progenitor cell proliferation, migration and VEGF production via PI 3-kinase/Akt signaling pathways. Acta Pharmacol Sin 27:685–693
126. Podolska K, Lipiec A, Hajdukiewicz K, Lubkowska H, Małecki M (2013) Sonic hedgehog stimulates the recruitment of endothelial progenitor cells. Med Wieku Rozwoj 17:151–156
127. Chinchilla P, Xiao L, Kazanietz MG, Riobo NA (2010) Hedgehog proteins activate pro-angiogenic responses in endothelial cells through non-canonical signaling pathways. Cell Cycle 9:570–579

128. He QW, Xia YP, Chen SC, Wang Y, Huang M, Huang Y, Li JY, Li YN, Gao Y, Mao L, Mei YW, Hu B (2013) Astrocyte-derived sonic hedgehog contributes to angiogenesis in brain microvascular endothelial cells via RhoA/ROCK pathway after oxygen-glucose deprivation. Mol Neurobiol 47:976–987
129. Renault MA, Roncalli J, Tongers J, Thorne T, Klyachko E, Misener S, Volpert OV, Mehta S, Burg A, Luedemann C, Qin G, Kishore R, Losordo DW (2010) Sonic hedgehog induces angiogenesis via Rho kinase-dependent signaling in endothelial cells. J Mol Cell Cardiol 49:490–498
130. Yao Q, Renault MA, Chapouly C, Vandierdonck S, Belloc I, Jaspar-Vinassa B, Daniel-Lamazière JM, Laffargue M, Merched A, Desgranges C, Gadeau AP (2014) Sonic hedgehog mediates a novel pathway of PDGF-BB-dependent vessel maturation. Blood. doi:10.1182/blood-2013-06-508689
131. Renault MA, Robbesyn F, Chapouly C, Yao Q, Vandierdonck S, Reynaud A, Belloc I, Traiffort E, Ruat M, Desgranges C, Gadeau AP (2013) Hedgehog-dependent regulation of angiogenesis and myogenesis is impaired in aged mice. Arterioscler Thromb Vasc Biol 33:2858–2866
132. Pola R, Ling LE, Silver M, Corbley MJ, Kearney M, Blake Pepinsky R, Shapiro R, Taylor FR, Baker DP, Asahara T, Isner JM (2001) The morphogen Sonic hedgehog is an indirect angiogenic agent upregulating two families of angiogenic growth factors. Nat Med 7:706–711
133. Straface G, Aprahamian T, Flex A, Gaetani E, Biscetti F, Smith RC, Pecorini G, Pola E, Angelini F, Stigliano E, Castellot JJ Jr, Losordo DW, Pola R (2009) Sonic hedgehog regulates angiogenesis and myogenesis during post-natal skeletal muscle regeneration. J Cell Mol Med 13:2424–2435
134. Preda MB, Valen G (2013) Evaluation of gene and cell-based therapies for cardiac regeneration. Curr Stem Cell Res Ther 8:304–312
135. Xiao Q, Hou N, Wang YP, He LS, He YH, Zhang GP, Yi Q, Liu SM, Chen MS, Luo JD (2012) Impaired sonic hedgehog pathway contributes to cardiac dysfunction in type 1 diabetic mice with myocardial infarction. Cardiovasc Res 95:507–516
136. Kicheva A, Bollenbach T, Wartlick O, Julicher F, Gonzalez-Gaitan M (2012) Investigating the principles of morphogen gradient formation: from tissues to cells. Curr Opin Genet Dev 22:527–532
137. Kusano KF, Pola R, Murayama T, Curry C, Kawamoto A, Iwakura A, Shintani S, Ii M, Asai J, Tkebuchava T, Thorne T, Takenaka H, Aikawa R, Goukassian D, von Samson P, Hamada H, Yoon YS, Silver M, Eaton E, Ma H, Heyd L, Kearney M, Munger W, Porter JA, Kishore R, Losordo DW (2005) Sonic hedgehog myocardial gene therapy: tissue repair through transient reconstitution of embryonic signaling. Nat Med 11:1197–1204
138. Johnson NR, Wang Y (2013) Controlled delivery of sonic hedgehog morphogen and its potential for cardiac repair. PLoS One 8:e63075
139. Mackie AR, Klyachko E, Thorne T, Schultz KM, Millay M, Ito A, Kamide CE, Liu T, Gupta R, Sahoo S, Misener S, Kishore R, Losordo DW (2012) Sonic hedgehog-modified human CD34+ cells preserve cardiac function after acute myocardial infarction. Circ Res 111:312–321
140. Agouni A, Mostefai HA, Porro C, Carusio N, Favre J, Richard V, Henrion D, Martínez MC, Andriantsitohaina R (2007) Sonic hedgehog carried by microparticles corrects endothelial injury through nitric oxide release. FASEB J 21:2735–2741
141. Soleti R, Benameur T, Porro C, Panaro MA, Andriantsitohaina R, Martínez MC (2009) Microparticles harboring Sonic Hedgehog promote angiogenesis through the upregulation of adhesion proteins and proangiogenic factors. Carcinogenesis 30:580–588
142. Benameur T, Soleti R, Porro C, Andriantsitohaina R, Martínez MC (2010) Microparticles carrying Sonic hedgehog favor neovascularization through the activation of nitric oxide pathway in mice. PLoS One 5:e12688

143. Benameur T, Tual-Chalot S, Andriantsitohaina R, Martínez MC (2010) PPARalpha is essential for microparticle-induced differentiation of mouse bone marrow-derived endothelial progenitor cells and angiogenesis. PLoS One 5:e12392
144. Soleti R, Martinez MC (2012) Sonic Hedgehog on microparticles and neovascularization. Vitam Horm 88:395–438
145. Albayati MA, Shearman CP (2013) Peripheral arterial disease and bypass surgery in the diabetic lower limb. Med Clin North Am 97:821–834
146. Ozdemir BA, Brownrigg J, Patel N, Jones KG, Thompson MM, Hinchliffe RJ (2013) Population-based screening for the prevention of lower extremity complications in diabetes. Diabetes Metab Res Rev 29:173–182
147. Jarajapu YP, Grant MB (2010) The promise of cell-based therapies for diabetic complications: challenges and solutions. Circ Res 106:854–869
148. Luo JD, Hu TP, Wang L, Chen MS, Liu SM, Chen AF (2009) Sonic hedgehog improves delayed wound healing via enhancing cutaneous nitric oxide function in diabetes. Am J Physiol Endocrinol Metab 297:E525–E531
149. Wang JM, Isenberg JS, Billiar TR, Chen AF (2013) Thrombospondin-1/CD36 pathway contributes to bone marrow-derived angiogenic cell dysfunction in type 1 diabetes via Sonic hedgehog pathway suppression. Am J Physiol Endocrinol Metabolism 305:E1464–E1472
150. Kusano KF, Allendoerfer KL, Munger W, Pola R, Bosch-Marce M, Kirchmair R, Yoon YS, Curry C, Silver M, Kearney M, Asahara T, Losordo DW (2004) Sonic hedgehog induces arteriogenesis in diabetic vasa nervorum and restores function in diabetic neuropathy. Arterioscler Thromb Vasc Biol 24:2102–2107
151. Asai J, Takenaka H, Kusano KF, Ii M, Luedemann C, Curry C, Eaton E, Iwakura A, Tsutsumi Y, Hamada H, Kishimoto S, Thorne T, Kishore R, Losordo DW (2006) Topical sonic hedgehog gene therapy accelerates wound healing in diabetes by enhancing endothelial progenitor cell-mediated microvascular remodeling. Circulation 113:2413–2424
152. Palladino M, Gatto I, Neri V, Straino S, Silver M, Tritarelli A, Piccioni A, Smith RC, Gaetani E, Losordo DW, Crea F, Capogrossi M, Pola R (2011) Pleiotropic beneficial effects of sonic hedgehog gene therapy in an experimental model of peripheral limb ischemia. Mol Ther 19:658–666
153. Renault MA, Vandierdonck S, Chapouly C, Yu Y, Qin G, Metras A, Couffinhal T, Losordo DW, Yao Q, Reynaud A, Jaspard-Vinassa B, Belloc I, Desgranges C, Gadeau AP (2013) Gli3 regulation of myogenesis is necessary for ischemia-induced angiogenesis. Circ Res 113:1148–1158
154. Palladino M, Gatto I, Neri V, Stigliano E, Smith RC, Pola E, Straino S, Gaetani E, Capogrossi M, Leone G, Hlatky L, Pola R (2012) Combined therapy with sonic hedgehog gene transfer and bone marrow-derived endothelial progenitor cells enhances angiogenesis and myogenesis in the ischemic skeletal muscle. J Vasc Res 49:425–431
155. Tual-Chalot S, Leonetti D, Andriantsitohaina R, Martinez MC (2011) Microvesicles: intracellular vectors of biological messages. Mol Interv 11:88–94
156. Baj-Krzyworzeka M, Majka M, Pratico D, Ratajczak J, Vilaire G, Kijowski J, Reca R, Janowska-Wieczorek A, Ratajczak MZ (2002) Platelet-derived microparticles stimulate proliferation, survival, adhesion, and chemotaxis of hematopoietic cells. Exp Hematol 30:450–459
157. Janowska-Wieczorek A, Majka M, Kijowski J, Baj-Krzyworzeka M, Reca R, Turner AR, Ratajczak J, Emerson SG, Kowalska MA, Ratajczak MZ (2001) Platelet-derived microparticles bind to hematopoietic stem/progenitor cells and enhance their engraftment. Blood 98:3143–3149
158. Azevedo LC, Pedro MA, Laurindo FR (2007) Circulating microparticles as therapeutic targets in cardiovascular diseases. Recent Patents Cardiovasc Drug Discov 2:41–51
159. Yang C, Mwaikambo BR, Zhu T, Gagnon C, Lafleur J, Seshadri S, Lachapelle P, Lavoie JC, Chemtob S, Hardy P (2008) Lymphocytic microparticles inhibit angiogenesis by stimulating oxidative stress and negatively regulating VEGF-induced pathways. Am J Physiol Regul Integr Comp Physiol 294:467–476

160. Yang C, Xiong W, Qiu Q, Shao Z, Hamel D, Tahiri H, Leclair G, Lachapelle P, Chemtob S, Hardy P (2012) Role of receptor-mediated endocytosis in the antiangiogenic effects of human T lymphoblastic cell-derived microparticles. Am J Physiol Regul Integr Comp Physiol 302: R941–R949
161. Tahiri H, Yang C, Duhamel F, Omri S, Picard E, Chemtob S, Hardy P (2013) p75 neurotrophin receptor participates in the choroidal antiangiogenic and apoptotic effects of T-lymphocyte-derived microparticles. Invest Ophthalmol Vis Sci 54:6084–6092
162. Mostefai HA, Agouni A, Carusio N, Mastronardi ML, Heymes C, Henrion D, Andriantsitohaina R, Martinez MC (2008) Phosphatidylinositol 3-kinase and xanthine oxidase regulate nitric oxide and reactive oxygen species productions by apoptotic lymphocyte microparticles in endothelial cells. J Immunol 180:5028–5035
163. Mostefai HA, Andriantsitohaina R, Martínez MC (2008) Plasma membrane microparticles in angiogenesis: role in ischemic diseases and in cancer. Physiol Res 57:311–320
164. Yang C, Gagnon C, Hou X, Hardy P (2010) Low density lipoprotein receptor mediates anti-VEGF effect of lymphocyte T-derived microparticles in Lewis lung carcinoma cells. Cancer Biol Ther 10:448–456
165. Waltenberger J (2009) VEGF resistance as a molecular basis to explain the angiogenesis paradox in diabetes mellitus. Biochem Soc Trans 37:1167–1170

Index

MIX
Papier aus verantwortungsvollen Quellen
Paper from responsible sources
FSC® C105338

If you have any concerns about our products,
you can contact us on
ProductSafety@springernature.com

In case Publisher is established outside the EU,
the EU authorized representative is:
Springer Nature Customer Service Center GmbH
Europaplatz 3, 69115 Heidelberg, Germany

Printed by Libri Plureos GmbH
in Hamburg, Germany